Mometrix
TEST PREPARATION

Virginia SOL Grade 3 Mathematics Secrets Study Guide

I0894322

Dear Future Exam Success Story

First of all, **THANK YOU** for purchasing Mometrix study materials!

Second, congratulations! You are one of the few determined test-takers who are committed to doing whatever it takes to excel on your exam. **You have come to the right place.** We developed these study materials with one goal in mind: to deliver you the information you need in a format that's concise and easy to use.

In addition to optimizing your guide for the content of the test, we've outlined our recommended steps for breaking down the preparation process into small, attainable goals so you can make sure you stay on track.

We've also analyzed the entire test-taking process, identifying the most common pitfalls and showing how you can overcome them and be ready for any curveball the test throws you.

Standardized testing is one of the biggest obstacles on your road to success, which only increases the importance of doing well in the high-pressure, high-stakes environment of test day. Your results on this test could have a significant impact on your future, and this guide provides the information and practical advice to help you achieve your full potential on test day.

Your success is our success

We would love to hear from you! If you would like to share the story of your exam success or if you have any questions or comments in regard to our products, please contact us at **800-673-8175** or **support@mometrix.com**.

Thanks again for your business and we wish you continued success!

Sincerely,
The Mometrix Test Preparation Team

Need more help? Check out our flashcards at:
http://MometrixFlashcards.com/SOL

Written and edited by the Mometrix Exam Secrets Test Prep Team
Printed in the United States of America

TABLE OF CONTENTS

Introduction

Thank you for purchasing this resource! You have made the choice to prepare yourself for a test that could have a huge impact on your future, and this guide is designed to help you be fully ready for test day. Obviously, it's important to have a solid understanding of the test material, but you also need to be prepared for the unique environment and stressors of the test, so that you can perform to the best of your abilities.

For this purpose, the first section that appears in this guide is the **Secret Keys**. We've devoted countless hours to meticulously researching what works and what doesn't, and we've boiled down our findings to the five most impactful steps you can take to improve your performance on the test. We start at the beginning with study planning and move through the preparation process, all the way to the testing strategies that will help you get the most out of what you know when you're finally sitting in front of the test.

We recommend that you start preparing for your test as far in advance as possible. However, if you've bought this guide as a last-minute study resource and only have a few days before your test, we recommend that you skip over the first two Secret Keys since they address a long-term study plan.

If you struggle with **test anxiety**, we strongly encourage you to check out our recommendations for how you can overcome it. Test anxiety is a formidable foe, but it can be beaten, and we want to make sure you have the tools you need to defeat it.

Secret Key #1 – Plan Big, Study Small

There's a lot riding on your performance. If you want to ace this test, you're going to need to keep your skills sharp and the material fresh in your mind. You need a plan that lets you review everything you need to know while still fitting in your schedule. We'll break this strategy down into three categories.

Information Organization

Start with the information you already have: the official test outline. From this, you can make a complete list of all the concepts you need to cover before the test. Organize these concepts into groups that can be studied together, and create a list of any related vocabulary you need to learn so you can brush up on any difficult terms. You'll want to keep this vocabulary list handy once you actually start studying since you may need to add to it along the way.

Time Management

Once you have your set of study concepts, decide how to spread them out over the time you have left before the test. Break your study plan into small, clear goals so you have a manageable task for each day and know exactly what you're doing. Then just focus on one small step at a time. When you manage your time this way, you don't need to spend hours at a time studying. Studying a small block of content for a short period each day helps you retain information better and avoid stressing over how much you have left to do. You can relax knowing that you have a plan to cover everything in time. In order for this strategy to be effective though, you have to start studying early and stick to your schedule. Avoid the exhaustion and futility that comes from last-minute cramming!

Study Environment

The environment you study in has a big impact on your learning. Studying in a coffee shop, while probably more enjoyable, is not likely to be as fruitful as studying in a quiet room. It's important to keep distractions to a minimum. You're only planning to study for a short block of time, so make the most of it. Don't pause to check your phone or get up to find a snack. It's also important to **avoid multitasking**. Research has consistently shown that multitasking will make your studying dramatically less effective. Your study area should also be comfortable and well-lit so you don't have the distraction of straining your eyes or sitting on an uncomfortable chair.

The time of day you study is also important. You want to be rested and alert. Don't wait until just before bedtime. Study when you'll be most likely to comprehend and remember. Even better, if you know what time of day your test will be, set that time aside for study. That way your brain will be used to working on that subject at that specific time and you'll have a better chance of recalling information.

Finally, it can be helpful to team up with others who are studying for the same test. Your actual studying should be done in as isolated an environment as possible, but the work of organizing the information and setting up the study plan can be divided up. In between study sessions, you can discuss with your teammates the concepts that you're all studying and quiz each other on the details. Just be sure that your teammates are as serious about the test as you are. If you find that your study time is being replaced with social time, you might need to find a new team.

Secret Key #2 – Make Your Studying Count

You're devoting a lot of time and effort to preparing for this test, so you want to be absolutely certain it will pay off. This means doing more than just reading the content and hoping you can remember it on test day. It's important to make every minute of study count. There are two main areas you can focus on to make your studying count.

Retention

It doesn't matter how much time you study if you can't remember the material. You need to make sure you are retaining the concepts. To check your retention of the information you're learning, try recalling it at later times with minimal prompting. Try carrying around flashcards and glance at one or two from time to time or ask a friend who's also studying for the test to quiz you.

To enhance your retention, look for ways to put the information into practice so that you can apply it rather than simply recalling it. If you're using the information in practical ways, it will be much easier to remember. Similarly, it helps to solidify a concept in your mind if you're not only reading it to yourself but also explaining it to someone else. Ask a friend to let you teach them about a concept you're a little shaky on (or speak aloud to an imaginary audience if necessary). As you try to summarize, define, give examples, and answer your friend's questions, you'll understand the concepts better and they will stay with you longer. Finally, step back for a big picture view and ask yourself how each piece of information fits with the whole subject. When you link the different concepts together and see them working together as a whole, it's easier to remember the individual components.

Finally, practice showing your work on any multi-step problems, even if you're just studying. Writing out each step you take to solve a problem will help solidify the process in your mind, and you'll be more likely to remember it during the test.

Modality

Modality simply refers to the means or method by which you study. Choosing a study modality that fits your own individual learning style is crucial. No two people learn best in exactly the same way, so it's important to know your strengths and use them to your advantage.

For example, if you learn best by visualization, focus on visualizing a concept in your mind and draw an image or a diagram. Try color-coding your notes, illustrating them, or creating symbols that will trigger your mind to recall a learned concept. If you learn best by hearing or discussing information, find a study partner who learns the same way or read aloud to yourself. Think about how to put the information in your own words. Imagine that you are giving a lecture on the topic and record yourself so you can listen to it later.

For any learning style, flashcards can be helpful. Organize the information so you can take advantage of spare moments to review. Underline key words or phrases. Use different colors for different categories. Mnemonic devices (such as creating a short list in which every item starts with the same letter) can also help with retention. Find what works best for you and use it to store the information in your mind most effectively and easily.

Secret Key #3 – Practice the Right Way

Your success on test day depends not only on how many hours you put into preparing, but also on whether you prepared the right way. It's good to check along the way to see if your studying is paying off. One of the most effective ways to do this is by taking practice tests to evaluate your progress. Practice tests are useful because they show exactly where you need to improve. Every time you take a practice test, pay special attention to these three groups of questions:

- The questions you got wrong
- The questions you had to guess on, even if you guessed right
- The questions you found difficult or slow to work through

This will show you exactly what your weak areas are, and where you need to devote more study time. Ask yourself why each of these questions gave you trouble. Was it because you didn't understand the material? Was it because you didn't remember the vocabulary? Do you need more repetitions on this type of question to build speed and confidence? Dig into those questions and figure out how you can strengthen your weak areas as you go back to review the material.

Additionally, many practice tests have a section explaining the answer choices. It can be tempting to read the explanation and think that you now have a good understanding of the concept. However, an explanation likely only covers part of the question's broader context. Even if the explanation makes perfect sense, **go back and investigate** every concept related to the question until you're positive you have a thorough understanding.

As you go along, keep in mind that the practice test is just that: practice. Memorizing these questions and answers will not be very helpful on the actual test because it is unlikely to have any of the same exact questions. If you only know the right answers to the sample questions, you won't be prepared for the real thing. **Study the concepts** until you understand them fully, and then you'll be able to answer any question that shows up on the test.

It's important to wait on the practice tests until you're ready. If you take a test on your first day of study, you may be overwhelmed by the amount of material covered and how much you need to learn. Work up to it gradually.

On test day, you'll need to be prepared for answering questions, managing your time, and using the test-taking strategies you've learned. It's a lot to balance, like a mental marathon that will have a big impact on your future. Like training for a marathon, you'll need to start slowly and work your way up. When test day arrives, you'll be ready.

Start with the strategies you've read in the first two Secret Keys—plan your course and study in the way that works best for you. If you have time, consider using multiple study resources to get different approaches to the same concepts. It can be helpful to see difficult concepts from more than one angle. Then find a good source for practice tests. Many times, the test website will suggest potential study resources or provide sample tests.

Practice Test Strategy

If you're able to find at least three practice tests, we recommend this strategy:

Untimed and Open-Book Practice

Take the first test with no time constraints and with your notes and study guide handy. Take your time and focus on applying the strategies you've learned.

Timed and Open-Book Practice

Take the second practice test open-book as well, but set a timer and practice pacing yourself to finish in time.

Timed and Closed-Book Practice

Take any other practice tests as if it were test day. Set a timer and put away your study materials. Sit at a table or desk in a quiet room, imagine yourself at the testing center, and answer questions as quickly and accurately as possible.

Keep repeating timed and closed-book tests on a regular basis until you run out of practice tests or it's time for the actual test. Your mind will be ready for the schedule and stress of test day, and you'll be able to focus on recalling the material you've learned.

Secret Key #4 – Pace Yourself

Once you're fully prepared for the material on the test, your biggest challenge on test day will be managing your time. Just knowing that the clock is ticking can make you panic even if you have plenty of time left. Work on pacing yourself so you can build confidence against the time constraints of the exam. Pacing is a difficult skill to master, especially in a high-pressure environment, so **practice is vital**.

Set time expectations for your pace based on how much time is available. For example, if a section has 60 questions and the time limit is 30 minutes, you know you have to average 30 seconds or less per question in order to answer them all. Although 30 seconds is the hard limit, set 25 seconds per question as your goal, so you reserve extra time to spend on harder questions. When you budget extra time for the harder questions, you no longer have any reason to stress when those questions take longer to answer.

Don't let this time expectation distract you from working through the test at a calm, steady pace, but keep it in mind so you don't spend too much time on any one question. Recognize that taking extra time on one question you don't understand may keep you from answering two that you do understand later in the test. If your time limit for a question is up and you're still not sure of the answer, mark it and move on, and come back to it later if the time and the test format allow. If the testing format doesn't allow you to return to earlier questions, just make an educated guess; then put it out of your mind and move on.

On the easier questions, be careful not to rush. It may seem wise to hurry through them so you have more time for the challenging ones, but it's not worth missing one if you know the concept and just didn't take the time to read the question fully. Work efficiently but make sure you understand the question and have looked at all of the answer choices, since more than one may seem right at first.

Even if you're paying attention to the time, you may find yourself a little behind at some point. You should speed up to get back on track, but do so wisely. Don't panic; just take a few seconds less on each question until you're caught up. Don't guess without thinking, but do look through the answer choices and eliminate any you know are wrong. If you can get down to two choices, it is often worthwhile to guess from those. Once you've chosen an answer, move on and don't dwell on any that you skipped or had to hurry through. If a question was taking too long, chances are it was one of the harder ones, so you weren't as likely to get it right anyway.

On the other hand, if you find yourself getting ahead of schedule, it may be beneficial to slow down a little. The more quickly you work, the more likely you are to make a careless mistake that will affect your score. You've budgeted time for each question, so don't be afraid to spend that time. Practice an efficient but careful pace to get the most out of the time you have.

Secret Key #5 – Have a Plan for Guessing

When you're taking the test, you may find yourself stuck on a question. Some of the answer choices seem better than others, but you don't see the one answer choice that is obviously correct. What do you do?

The scenario described above is very common, yet most test takers have not effectively prepared for it. Developing and practicing a plan for guessing may be one of the single most effective uses of your time as you get ready for the exam.

In developing your plan for guessing, there are three questions to address:

- When should you start the guessing process?
- How should you narrow down the choices?
- Which answer should you choose?

When to Start the Guessing Process

Unless your plan for guessing is to select C every time (which, despite its merits, is not what we recommend), you need to leave yourself enough time to apply your answer elimination strategies. Since you have a limited amount of time for each question, that means that if you're going to give yourself the best shot at guessing correctly, you have to decide quickly whether or not you will guess.

Of course, the best-case scenario is that you don't have to guess at all, so first, see if you can answer the question based on your knowledge of the subject and basic reasoning skills. Focus on the key words in the question and try to jog your memory of related topics. Give yourself a chance to bring the knowledge to mind, but once you realize that you don't have (or you can't access) the knowledge you need to answer the question, it's time to start the guessing process.

It's almost always better to start the guessing process too early than too late. It only takes a few seconds to remember something and answer the question from knowledge. Carefully eliminating wrong answer choices takes longer. Plus, going through the process of eliminating answer choices can actually help jog your memory.

Summary: Start the guessing process as soon as you decide that you can't answer the question based on your knowledge.

How to Narrow Down the Choices

The next chapter in this book (**Test-Taking Strategies**) includes a wide range of strategies for how to approach questions and how to look for answer choices to eliminate. You will definitely want to read those carefully, practice them, and figure out which ones work best for you. Here though, we're going to address a mindset rather than a particular strategy.

Your odds of guessing an answer correctly depend on how many options you are choosing from.

Number of options left	5	4	3	2	1
Odds of guessing correctly	20%	25%	33%	50%	100%

You can see from this chart just how valuable it is to be able to eliminate incorrect answers and make an educated guess, but there are two things that many test takers do that cause them to miss out on the benefits of guessing:

- Accidentally eliminating the correct answer
- Selecting an answer based on an impression

We'll look at the first one here, and the second one in the next section.

To avoid accidentally eliminating the correct answer, we recommend a thought exercise called **the $5 challenge**. In this challenge, you only eliminate an answer choice from contention if you are willing to bet $5 on it being wrong. Why $5? Five dollars is a small but not insignificant amount of money. It's an amount you could afford to lose but wouldn't want to throw away. And while losing $5 once might not hurt too much, doing it twenty times will set you back $100. In the same way, each small decision you make—eliminating a choice here, guessing on a question there—won't by itself impact your score very much, but when you put them all together, they can make a big difference. By holding each answer choice elimination decision to a higher standard, you can reduce the risk of accidentally eliminating the correct answer.

The $5 challenge can also be applied in a positive sense: If you are willing to bet $5 that an answer choice *is* correct, go ahead and mark it as correct.

Summary: Only eliminate an answer choice if you are willing to bet $5 that it is wrong.

Which Answer to Choose

You're taking the test. You've run into a hard question and decided you'll have to guess. You've eliminated all the answer choices you're willing to bet $5 on. Now you have to pick an answer. Why do we even need to talk about this? Why can't you just pick whichever one you feel like when the time comes?

The answer to these questions is that if you don't come into the test with a plan, you'll rely on your impression to select an answer choice, and if you do that, you risk falling into a trap. The test writers know that everyone who takes their test will be guessing on some of the questions, so they intentionally write wrong answer choices to seem plausible. You still have to pick an answer though, and if the wrong answer choices are designed to look right, how can you ever be sure that you're not falling for their trap? The best solution we've found to this dilemma is to take the decision out of your hands entirely. Here is the process we recommend:

Once you've eliminated any choices that you are confident (willing to bet $5) are wrong, select the first remaining choice as your answer.

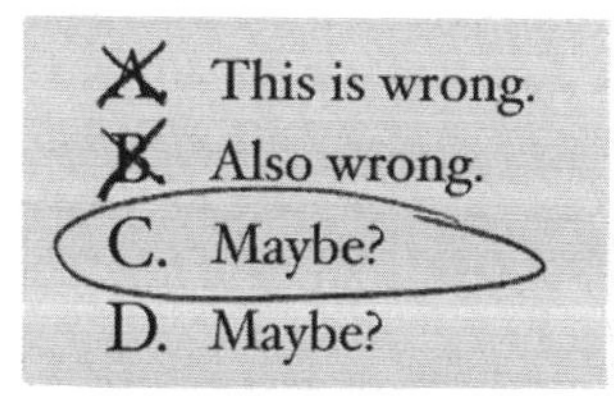

Whether you choose to select the first remaining choice, the second, or the last, the important thing is that you use some preselected standard. Using this approach guarantees that you will not be enticed into selecting an answer choice that looks right, because you are not basing your decision on how the answer choices look.

This is not meant to make you question your knowledge. Instead, it is to help you recognize the difference between your knowledge and your impressions. There's a huge difference between thinking an answer is right because of what you know, and thinking an answer is right because it looks or sounds like it should be right.

Summary: To ensure that your selection is appropriately random, make a predetermined selection from among all answer choices you have not eliminated.

Test-Taking Strategies

This section contains a list of test-taking strategies that you may find helpful as you work through the test. By taking what you know and applying logical thought, you can maximize your chances of answering any question correctly!

It is very important to realize that every question is different and every person is different: no single strategy will work on every question, and no single strategy will work for every person. That's why we've included all of them here, so you can try them out and determine which ones work best for different types of questions and which ones work best for you.

Question Strategies

✓ Read Carefully

Read the question and the answer choices carefully. Don't miss the question because you misread the terms. You have plenty of time to read each question thoroughly and make sure you understand what is being asked. Yet a happy medium must be attained, so don't waste too much time. You must read carefully and efficiently.

✓ Contextual Clues

Look for contextual clues. If the question includes a word you are not familiar with, look at the immediate context for some indication of what the word might mean. Contextual clues can often give you all the information you need to decipher the meaning of an unfamiliar word. Even if you can't determine the meaning, you may be able to narrow down the possibilities enough to make a solid guess at the answer to the question.

✓ Prefixes

If you're having trouble with a word in the question or answer choices, try dissecting it. Take advantage of every clue that the word might include. Prefixes can be a huge help. Usually, they allow you to determine a basic meaning. *Pre-* means before, *post-* means after, *pro-* is positive, *de-* is negative. From prefixes, you can get an idea of the general meaning of the word and try to put it into context.

✓ Hedge Words

Watch out for critical hedge words, such as *likely, may, can, often, almost, mostly, usually, generally, rarely,* and *sometimes.* Question writers insert these hedge phrases to cover every possibility. Often an answer choice will be wrong simply because it leaves no room for exception. Be on guard for answer choices that have definitive words such as *exactly* and *always.*

✓ Switchback Words

Stay alert for *switchbacks.* These are the words and phrases frequently used to alert you to shifts in thought. The most common switchback words are *but, although,* and *however.* Others include *nevertheless, on the other hand, even though, while, in spite of, despite,* and *regardless of.* Switchback words are important to catch because they can change the direction of the question or an answer choice.

✓ FACE VALUE

When in doubt, use common sense. Accept the situation in the problem at face value. Don't read too much into it. These problems will not require you to make wild assumptions. If you have to go beyond creativity and warp time or space in order to have an answer choice fit the question, then you should move on and consider the other answer choices. These are normal problems rooted in reality. The applicable relationship or explanation may not be readily apparent, but it is there for you to figure out. Use your common sense to interpret anything that isn't clear.

Answer Choice Strategies

✓ ANSWER SELECTION

The most thorough way to pick an answer choice is to identify and eliminate wrong answers until only one is left, then confirm it is the correct answer. Sometimes an answer choice may immediately seem right, but be careful. The test writers will usually put more than one reasonable answer choice on each question, so take a second to read all of them and make sure that the other choices are not equally obvious. As long as you have time left, it is better to read every answer choice than to pick the first one that looks right without checking the others.

✓ ANSWER CHOICE FAMILIES

An answer choice family consists of two (in rare cases, three) answer choices that are very similar in construction and cannot all be true at the same time. If you see two answer choices that are direct opposites or parallels, one of them is usually the correct answer. For instance, if one answer choice says that quantity *x* increases and another either says that quantity *x* decreases (opposite) or says that quantity *y* increases (parallel), then those answer choices would fall into the same family. An answer choice that doesn't match the construction of the answer choice family is more likely to be incorrect. Most questions will not have answer choice families, but when they do appear, you should be prepared to recognize them.

✓ ELIMINATE ANSWERS

Eliminate answer choices as soon as you realize they are wrong, but make sure you consider all possibilities. If you are eliminating answer choices and realize that the last one you are left with is also wrong, don't panic. Start over and consider each choice again. There may be something you missed the first time that you will realize on the second pass.

✓ AVOID FACT TRAPS

Don't be distracted by an answer choice that is factually true but doesn't answer the question. You are looking for the choice that answers the question. Stay focused on what the question is asking for so you don't accidentally pick an answer that is true but incorrect. Always go back to the question and make sure the answer choice you've selected actually answers the question and is not merely a true statement.

✓ EXTREME STATEMENTS

In general, you should avoid answers that put forth extreme actions as standard practice or proclaim controversial ideas as established fact. An answer choice that states the "process should be used in certain situations, if..." is much more likely to be correct than one that states the "process should be discontinued completely." The first is a calm rational statement and doesn't even make a definitive, uncompromising stance, using a hedge word *if* to provide wiggle room, whereas the second choice is far more extreme.

☑ Benchmark

As you read through the answer choices and you come across one that seems to answer the question well, mentally select that answer choice. This is not your final answer, but it's the one that will help you evaluate the other answer choices. The one that you selected is your benchmark or standard for judging each of the other answer choices. Every other answer choice must be compared to your benchmark. That choice is correct until proven otherwise by another answer choice beating it. If you find a better answer, then that one becomes your new benchmark. Once you've decided that no other choice answers the question as well as your benchmark, you have your final answer.

☑ Predict the Answer

Before you even start looking at the answer choices, it is often best to try to predict the answer. When you come up with the answer on your own, it is easier to avoid distractions and traps because you will know exactly what to look for. The right answer choice is unlikely to be word-for-word what you came up with, but it should be a close match. Even if you are confident that you have the right answer, you should still take the time to read each option before moving on.

General Strategies

☑ Tough Questions

If you are stumped on a problem or it appears too hard or too difficult, don't waste time. Move on! Remember though, if you can quickly check for obviously incorrect answer choices, your chances of guessing correctly are greatly improved. Before you completely give up, at least try to knock out a couple of possible answers. Eliminate what you can and then guess at the remaining answer choices before moving on.

☑ Check Your Work

Since you will probably not know every term listed and the answer to every question, it is important that you get credit for the ones that you do know. Don't miss any questions through careless mistakes. If at all possible, try to take a second to look back over your answer selection and make sure you've selected the correct answer choice and haven't made a costly careless mistake (such as marking an answer choice that you didn't mean to mark). This quick double check should more than pay for itself in caught mistakes for the time it costs.

☑ Pace Yourself

It's easy to be overwhelmed when you're looking at a page full of questions; your mind is confused and full of random thoughts, and the clock is ticking down faster than you would like. Calm down and maintain the pace that you have set for yourself. Especially as you get down to the last few minutes of the test, don't let the small numbers on the clock make you panic. As long as you are on track by monitoring your pace, you are guaranteed to have time for each question.

☑ Don't Rush

It is very easy to make errors when you are in a hurry. Maintaining a fast pace in answering questions is pointless if it makes you miss questions that you would have gotten right otherwise. Test writers like to include distracting information and wrong answers that seem right. Taking a little extra time to avoid careless mistakes can make all the difference in your test score. Find a pace that allows you to be confident in the answers that you select.

✓ Keep Moving

Panicking will not help you pass the test, so do your best to stay calm and keep moving. Taking deep breaths and going through the answer elimination steps you practiced can help to break through a stress barrier and keep your pace.

Final Notes

The combination of a solid foundation of content knowledge and the confidence that comes from practicing your plan for applying that knowledge is the key to maximizing your performance on test day. As your foundation of content knowledge is built up and strengthened, you'll find that the strategies included in this chapter become more and more effective in helping you quickly sift through the distractions and traps of the test to isolate the correct answer.

Now that you're preparing to move forward into the test content chapters of this book, be sure to keep your goal in mind. As you read, think about how you will be able to apply this information on the test. If you've already seen sample questions for the test and you have an idea of the question format and style, try to come up with questions of your own that you can answer based on what you're reading. This will give you valuable practice applying your knowledge in the same ways you can expect to on test day.

Good luck and good studying!

Number and Number Sense

Place Value

Numbers can be broken apart and composed based on their **place value**. For example, the number 129,548 can be expressed in **expanded form** as 100,000 + 20,000 + 9,000 + 500 + 40 + 8. You should also be comfortable going from expanded form to standard form. For example, what number is represented by 30,000 + 400 + 80 + 2? Notice 3 groups of ten-thousands, 0 groups of thousands, 4 groups of hundreds, 8 groups of tens, and 2 groups of ones. This becomes 30,482.

Sometimes you will be asked to compose and decompose numbers that are presented in **picture** or **model** form. This usually includes base-10 blocks. For example, consider the model below. How many groups of thousands, hundreds, tens, and ones are there? There is 1 group of 1,000, 2 groups of 100, 4 groups of 10, and 7 ones. This forms the number 1,247.

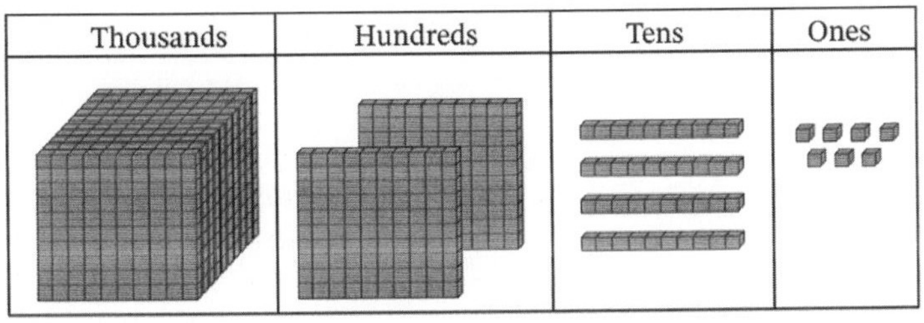

EXAMPLE

Decompose the number 452,837 into expanded form.

The number 452,837 can be decomposed based on its place value. The 4 in the hundred-thousands place represents 400,000. The 5 in the ten-thousands place represents 50,000. The 2 in the thousands place represents 2,000. The 8 in the hundreds place represents 800. The 3 in the tens place represents 30. The 7 in the ones place represents 7. Altogether, 452,837 in expanded form is 400,000 + 50,000 + 2,000 + 800 + 30 + 7.

EXAMPLE

What is the combined value of all of the base ten blocks below? Express the number in expanded form.

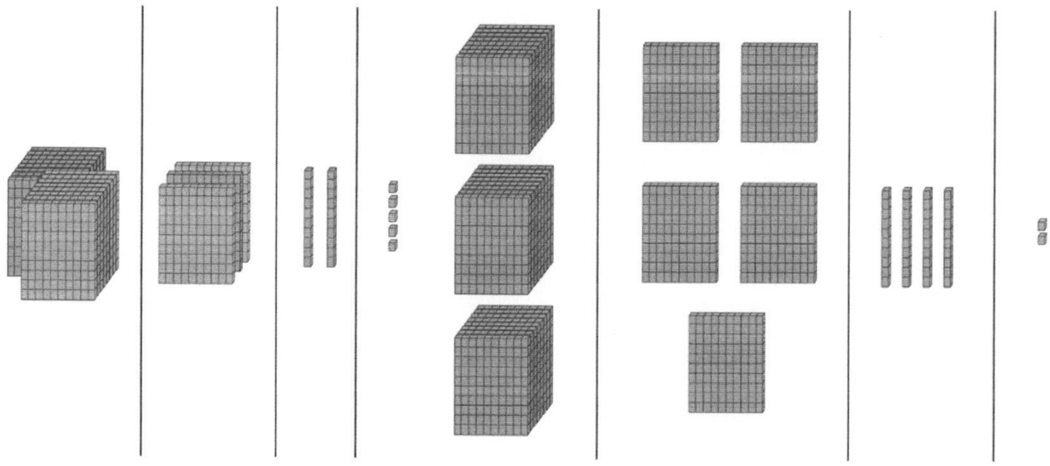

There are 5 blocks of 1,000. There are 8 plates of 100. There are 6 rods of 10. There are 7 cubes of 1. $5{,}000 + 800 + 60 + 7$ represents the amount in expanded form. In standard form the value is 5,867.

THE BASE-10 NUMBER SYSTEM

Consider the numbers 3, 30, 300, and 3,000. What happens from one number to the next? Each number is **10 times** the previous number. The great thing about the **base-10 number system** is the **consistent pattern** regarding place value. As you move from left to right, place value increases by a factor of 10. As you noticed in the previous example, 30 is 10 times as large as 3, 300 is 10 times as large as 30, 3,000 is 10 times as large as 300. This concept can be applied when you jump from a small place value to a large place value. For example, consider 45 and 450,000. How do we get from 45 to 450,000?

$$45 \times 10 = 450$$

$$45 \times 100 = 4{,}500$$

$$45 \times 1{,}000 = 45{,}000$$

$$45 \times 10{,}000 = 450{,}000$$

The important thing to remember is that each place value increases by a factor of 10. Another way to think about this is $45 \times 10 \times 10 \times 10 \times 10 \ = \ 450{,}000$.

Review Video: Number Place Value
Visit mometrix.com/academy and enter code: 205433

Rounding

There are many ways to round numbers. For example, when the number 89 is rounded to the nearest ten it becomes 90. When 89 is rounded to the nearest hundred it becomes 100.

A **number line** is a great tool to use when **rounding** numbers.

When rounding to the nearest ten, set up a number line with intervals of 10. Plot the number on the number line and look for the multiple of 10 that the number is closest to. For example, notice how 18 is closer to 20 than to 10. This means that 18 rounded to the nearest ten is 20.

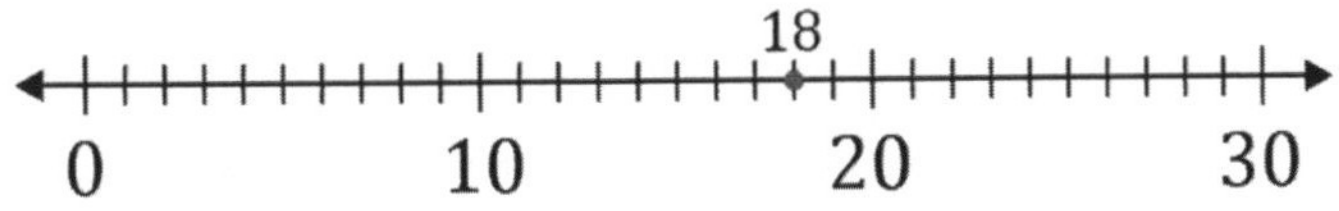

The same process works for numbers with larger place values. Consider the number 341. Create a number line with intervals of 100. Plot 341 on the number line. Notice that it is closer to 300 than 400. This means 341 rounded to the nearest hundred is 300.

Let's look at one more. Consider the number 1,627. Plot the point on a number line that has intervals of 1,000. Notice how 1,627 is closer to 2,000 than 1,000. This means that 1,627 rounded to the nearest thousand, which is 2,000.

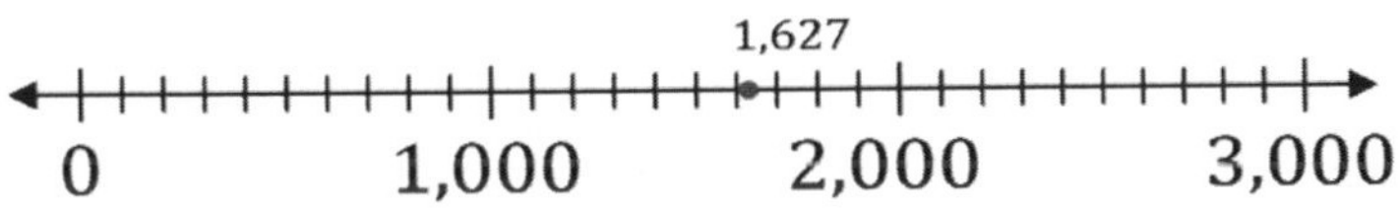

Comparing Numbers

Two important symbols are the **inequality symbols** for **greater than** (>) and **less than** (<). For example, 45 > 35 means 45 is greater than 35, and 26 < 62 means 26 is less than 62. We can now build on this skill to compare and order whole numbers up to 100,000. The important thing to remember is place value. Remember, when comparing numbers, start by comparing the **farthest left digit**. For example, when comparing 452,987 and 455,798, start by comparing the hundred-thousands place.

452,987 and 455,798

In this case, both values are 4. Since they are the same, move to the ten-thousands place.

452,987 and 455,798

Both values are both 5, so move to the thousands place.

452,987 and 455,798

2 is less than 5, so 452,987 is less than 455,798.

This can be expressed using an inequality symbol: 452,987 < 455,798.

Example

Order the three values from least to greatest using the symbols < and >:

65,033; 65,300; 65,030.

When comparing numbers, start with the farthest left digit and move right until the digits differ. All three numbers have the same digit in the ten-thousands place (6). Move right. All three numbers have the same digit in the thousands place (5). Move right. 65,300 has the largest digit in the hundreds place (3), therefore it is the largest number. Move right. Two numbers remain. Both 65,030 and 65,033 have the same digit in the tens place (3). Move right. 65,033 has a larger digit in the ones place, therefore it is larger than 65,030. The numbers from least to greatest are 65,030 < 65,033 < 65,300.

Example

Use an inequality symbol to compare 905,532 and 905,600.

Compare the numbers by comparing each digit from left to right. The digits in the hundred-thousands, ten-thousands, and thousands place are all the same (905). Keep moving right. The hundreds digits differ. One says 6 and one says 5. This means that 905,600 is larger than 905,532. 905,600 > 905,532, or 905,532 < 905,600.

Fractions

Values that are greater than 0 but less than or equal to 1 can be represented as a **fraction**. For example, the pizza, watermelon, and lemon are each sliced into equal pieces. The pizza has 6 equal pieces, the watermelon has 4 equal pieces, and the lemon has 8 equal pieces. These values represent the denominator of a fraction. Remember, the **denominator** is the bottom number in a fraction, which represents the **total number of pieces**.

$\frac{?}{6}$ $\frac{?}{4}$ $\frac{?}{8}$

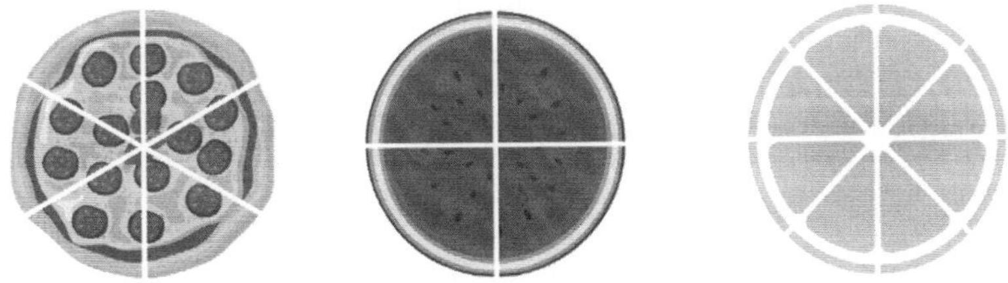

If 3 slices of each food are eaten, 3 would be the **numerator**.

$\frac{3}{6}$ of the pizza is eaten. $\frac{3}{4}$ of the watermelon is eaten. $\frac{3}{8}$ of the lemon is eaten.

Fractions can also be modeled using number lines or **fraction strips**.

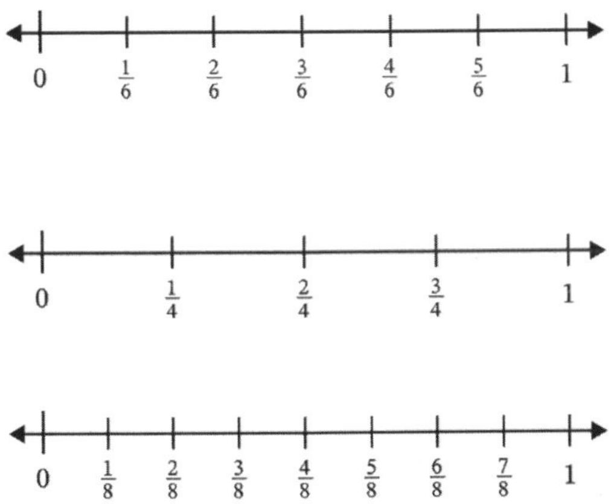

$\frac{1}{6}$	$\frac{1}{6}$	$\frac{1}{6}$	$\frac{1}{6}$	$\frac{1}{6}$	$\frac{1}{6}$

$\frac{1}{4}$	$\frac{1}{4}$	$\frac{1}{4}$	$\frac{1}{4}$

$\frac{1}{8}$	$\frac{1}{8}$	$\frac{1}{8}$	$\frac{1}{8}$	$\frac{1}{8}$	$\frac{1}{8}$	$\frac{1}{8}$	$\frac{1}{8}$

Fractions and Number Lines

Fractions are often displayed using models, fraction strips, pictures, or number lines. When determining a fraction from a number line, check to see how many **parts** there are within **one whole**. For example, consider the number line below. Notice how there are 4 **equal parts** from 0 to 1. This means each hash mark represents 1 out of 4, or $\frac{1}{4}$. Count up by $\frac{1}{4}$ to determine the missing value. $\frac{1}{4}, \frac{2}{4}, \frac{3}{4}$... The missing value is $\frac{3}{4}$.

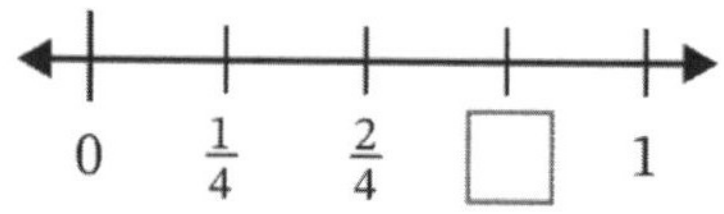

Consider the number line below. To identify the fraction that the green dot represents, first figure out the interval. The **interval** is essentially how many equal parts are shown to be within one whole number. In this case, there are 10 equal parts from 0 to 1. This means the interval is tenths. Starting from 0, or $\frac{0}{10}$, each hop to the right increases by $\frac{1}{10}$. The dot represents 8 hops, so $\frac{8}{10}$.

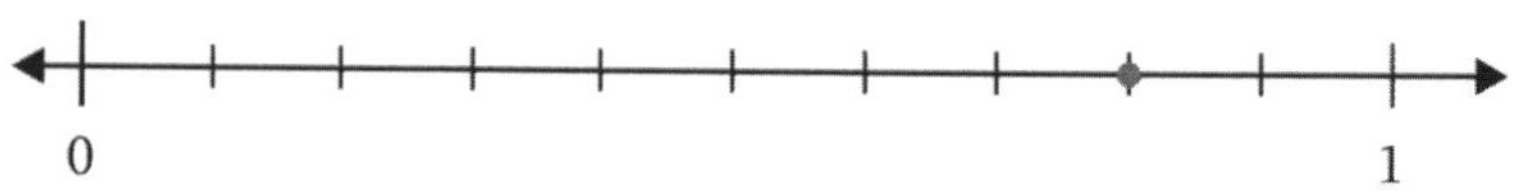

Unit Fractions

A **unit fraction** represents **one part** out of the whole. For example, $\frac{1}{2}, \frac{1}{4}, \frac{1}{6}$ and $\frac{1}{10}$ are all considered unit fractions. If the **numerator** is **1**, then the fraction is considered a unit fraction. Unit fractions can be shown in a model. For example, if a pan of brownies is cut into 8 equal pieces and 1 brownie is eaten, the consumed brownie represents the unit fraction $\frac{1}{8}$. The numerator is 1, and the denominator represents the total number of parts.

Composing and Decomposing Fractions

Fractions can be thought of as a **collection** of unit fractions. For example, the fraction $\frac{4}{5}$ is composed of $\frac{1}{5}+\frac{1}{5}+\frac{1}{5}+\frac{1}{5}$. The opposite of **composing** a fraction is **decomposing** a fraction. When a fraction is **broken up** into the sum of its unit fractions, it has been decomposed. For example, $\frac{6}{8}$ can be decomposed into $\frac{1}{8}+\frac{1}{8}+\frac{1}{8}+\frac{1}{8}+\frac{1}{8}+\frac{1}{8}$.

Using Models to Solve Problems Involving Fractions

Fractions represent part of a whole. One helpful way to see this is to draw a model. For example, if a cake is sliced into 8 equal slices, we can model the scenario by drawing lines to divide the cake into 8 slices. If 4 slices are eaten, then $\frac{4}{8}$ remain.

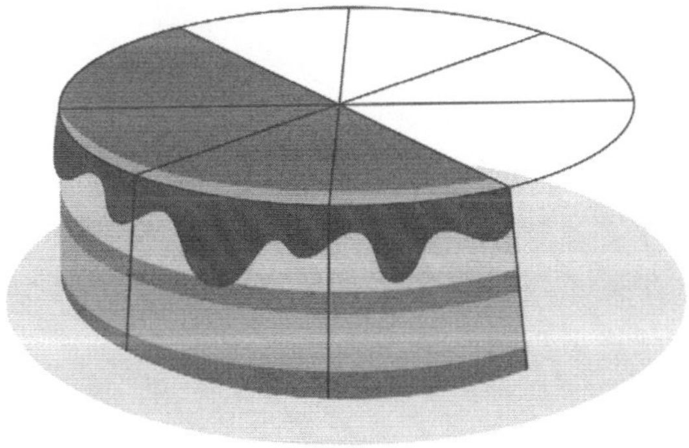

What if a cake had 10 slices, and each person wanted 2 slices? How many people could this cake feed?

Draw a quick picture of a cake sliced into 10 pieces. How many groups of 2 slices are there? A cake with 10 slices has 5 groups of 2. This cake would feed 5 people.

Drawing a quick sketch is often a helpful strategy for problems involving equal pieces.

EXAMPLE

Use the fraction strips to identify an equivalent fraction to $\frac{2}{3}$.

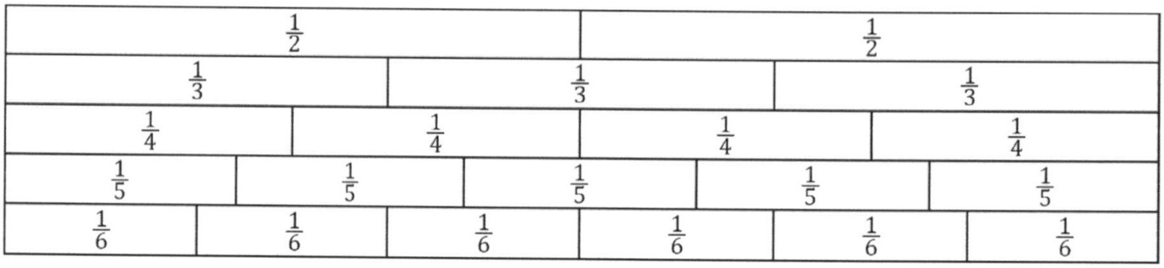

Locate the fraction strips for $\frac{1}{3}$. Move to the right to locate $\frac{2}{3}$. Notice how $\frac{2}{3}$ is the same length as $\frac{4}{6}$. The fractions $\frac{2}{3}$ and $\frac{4}{6}$ are considered equivalent fractions. $\frac{4}{6}$ is the correct answer.

EXAMPLE

Use the fraction number lines to identify three fractions that are equivalent to $\frac{4}{8}$.

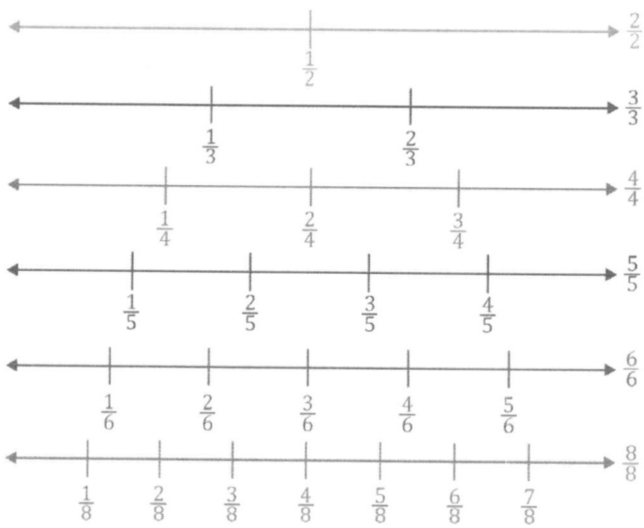

Equivalent fractions represent the same distance from 0 on a number line. Locate the fraction $\frac{4}{8}$. Compare this distance with the other number lines. Notice how the red, blue, and orange number lines have values at the same distance. The fractions $\frac{1}{2}, \frac{2}{4}$, and $\frac{3}{6}$ are the same distance from 0 as $\frac{4}{8}$. This means $\frac{1}{2}, \frac{2}{4}, \frac{3}{6}$, and $\frac{4}{8}$ are all equivalent fractions. The answer is $\frac{1}{2}, \frac{2}{4}, \frac{3}{6}$.

EQUIVALENT FRACTIONS

Equivalent fractions are fractions that represent the **same amount**. For example, $\frac{1}{2}$ of a pizza is the same amount as $\frac{2}{4}$ of a pizza. It can be helpful to show equivalent fractions on a number line.

Consider the number lines below. Each number line extends from 0 to 1, and all of the number lines are the same length.

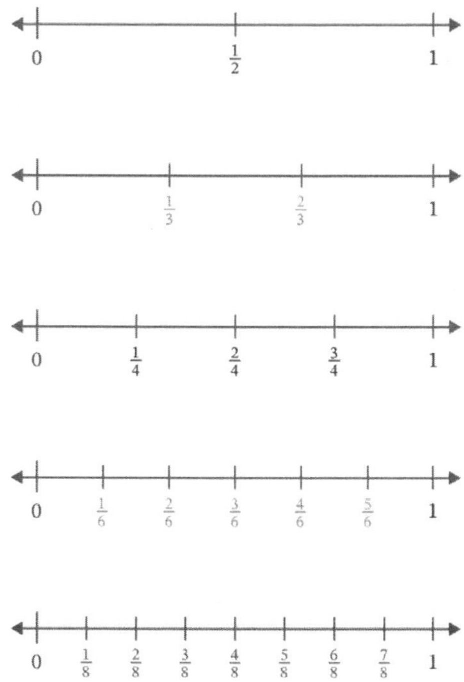

Notice how the distance at $\frac{1}{2}$ is the same distance as $\frac{2}{4}$, $\frac{3}{6}$, and $\frac{4}{8}$. These are all considered equivalent fractions because they all represent the same amount. $\frac{1}{2}$, $\frac{2}{4}$, $\frac{3}{6}$, and $\frac{4}{8}$ are all the same distance from 0 on the number line.

The same concept can be seen using a model. The circle below is made up of 3 equal parts. One part is shaded. This shaded part represents $\frac{1}{3}$. What is another fraction equivalent to $\frac{1}{3}$?

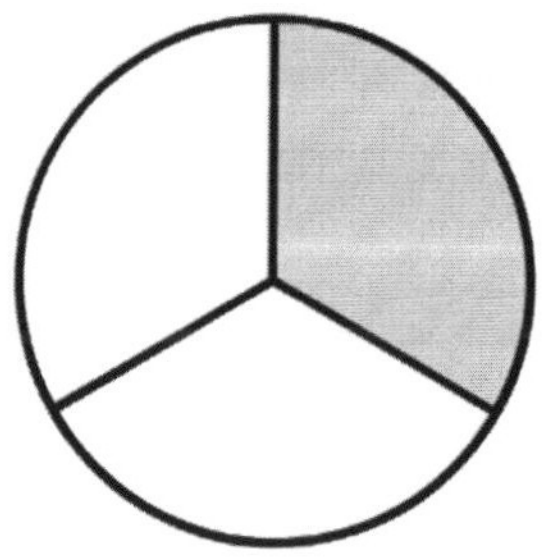

Instead of dividing the circle into 3 equal parts, divide it into 6 equal parts. Notice how 1 out of 3 becomes 2 out of 6. The fraction $\frac{2}{6}$ is equivalent to $\frac{1}{3}$.

Remember, the term **equivalent** means the same amount. It can be helpful to find equivalent fractions using a number line. Fractions that are the same distance from 0 on a number line are considered equivalent fractions. Notice how the two red points on the number lines below are the same distance from 0. The fractions $\frac{1}{3}$ and $\frac{2}{6}$ represent the same distance, therefore they are equivalent. Are there any other pairs of equivalent fractions visible on the number lines below?

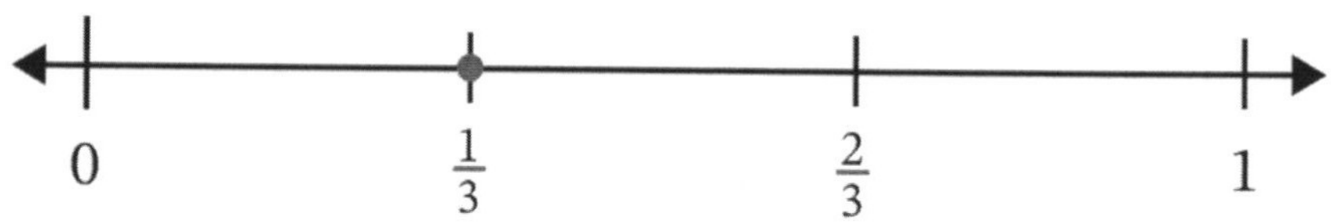

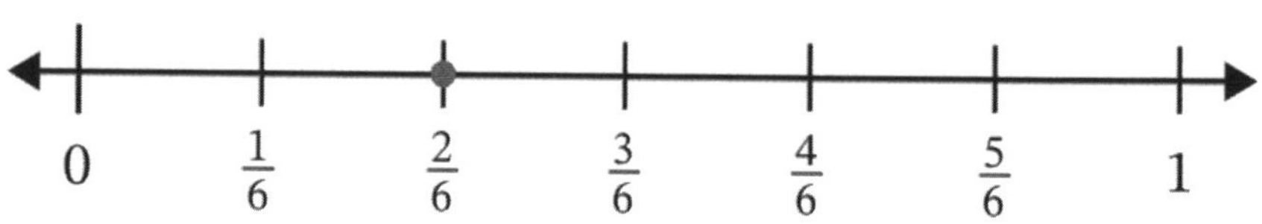

Yes! $\frac{2}{3}$ and $\frac{4}{6}$ are the same distance from 0. $\frac{2}{3}$ and $\frac{4}{6}$ are equivalent fractions.

Equivalent fractions can also be determined using an area model. Consider the fraction $\frac{1}{2}$. The area model shows a rectangle with 2 equal pieces, and 1 part shaded. This represents $\frac{1}{2}$. What is another way to express $\frac{1}{2}$? In other words, what is an equivalent fraction?

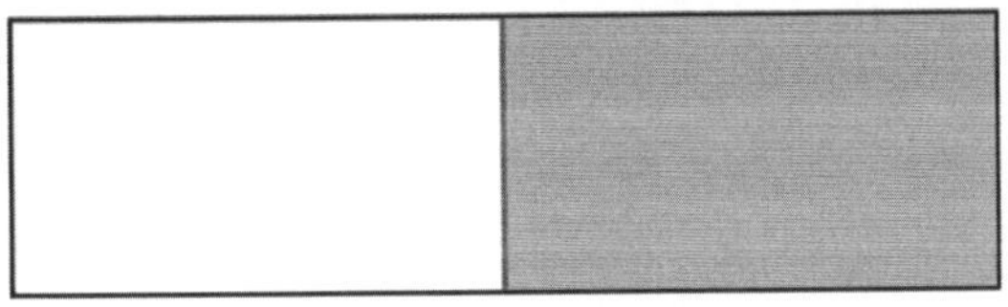

What if the same rectangle was sliced into 8 equal pieces? How many pieces would be shaded?

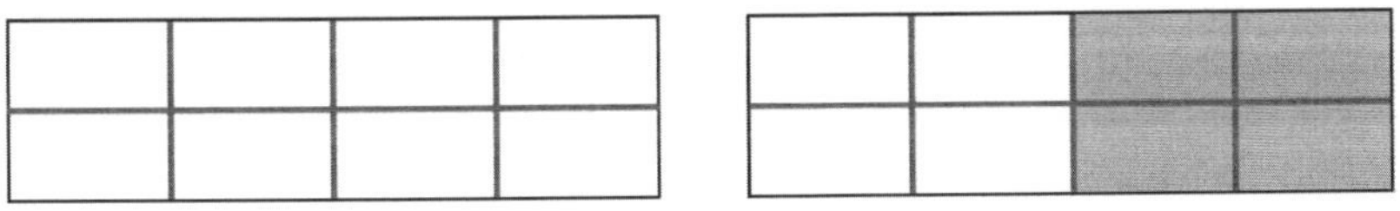

When the model is sliced into 8 equal pieces, 4 parts are shaded. $\frac{1}{2}$ and $\frac{4}{8}$ are considered equivalent fractions.

Comparing Fractions

Many fractions can be compared using logic. For example, consider the unit fractions $\frac{1}{4}$ and $\frac{1}{3}$. These fractions could be compared using a number line or by drawing a picture. However, these tools are not always necessary.

Both fractions are unit fractions, so they represent one part out of the total. $\frac{1}{4}$ represents one of the fourths. $\frac{1}{3}$ represents one of the thirds. Thirds are larger than fourths, so $\frac{1}{3}$ is larger than $\frac{1}{4}$.

The numerators are the same, and thirds are larger than fourths: $\frac{1}{3} > \frac{1}{4}$.

Now consider two fractions that have the same denominator. Compare the two fractions $\frac{2}{4}$ and $\frac{3}{4}$. Again, a picture or number line is not always necessary.

Both fractions have 4 in the denominator. This means both fractions have a whole of 4. However, 2 parts out of 4 is less than 3 parts out of 4. Think: 2 pieces are less than 3 pieces, so $\frac{2}{4}$ is less than $\frac{3}{4}$: $\frac{2}{4} < \frac{3}{4}$.

Example

Use an inequality symbol to compare the two fractions $\frac{4}{5}$ and $\frac{4}{6}$.

The numerator of a fraction represents the "parts". The denominator of a fraction represents the "whole". The fractions $\frac{4}{5}$ and $\frac{4}{6}$ have the same number of parts but a different number of total parts.

Fifths are larger than sixths. Therefore, 4 out of 5 is more than 4 out of 6. Think about it this way: 4 large pieces is more than 4 small pieces.

Notice how 4 pieces out of 5 is more than 4 pieces out of 6.

$$\frac{4}{5} > \frac{4}{6}$$

EXAMPLE

Use an inequality symbol to compare the two fractions: $\frac{3}{8}$ and $\frac{2}{8}$.

Both fractions have the same denominator. $\frac{3}{8}$ and $\frac{2}{8}$ both have 8 total parts. 3 out of 8 parts is more than 2 out of 8 parts. This means that $\frac{3}{8}$ is greater than $\frac{2}{8}$.

Notice how 3 pieces out of 8 is more than 2 pieces out of 8.

$$\frac{3}{8} > \frac{2}{8}$$

Computation and Estimation

Solving Problems

Estimating Solutions

Estimating sums and differences is a way to come up with an answer that is close to the exact answer. For example, if you run 9 miles each week for 7 weeks, how many total miles have you run? 9 rounded to the nearest 10 is 10. 7 groups of 10 is 70. You have run about 70 miles. This is close to the actual answer, which is 63 miles.

Compatible numbers make addition and subtraction easier. **Compatible numbers** are numbers that make addition or subtraction easier to do, and often they end in a 0 or 5. Instead of calculating $257 - 49$, consider using compatible numbers. 257 is close to 250, and 49 is close to 50, and $250 - 50$ are compatible because they can be quickly subtracted: $250 - 50 = 200$. This is close to the actual answer of $257 - 49 = 208$.

Rounding

Rounding is another strategy that saves time when adding and subtracting. If the digit to the right is 5 or greater, then the rounded number will **round** up. If the digit to the right is less than 5, then the rounded number stays the same.

Round 248 to the nearest 10.

248: 4 is in the tens place. The digit to the right of the 4 will indicate whether the 4 stays the same or rounds up to 5. To the right of 4 there is an 8. This means the 4 will round up to 5. 248 becomes 250 when rounded to the nearest 10.

Round 248 to the nearest 100.

248: 2 is in the hundreds place. The digit to the right of the 2 will indicate whether the 2 stays the same or rounds up to 3. To the right of 2 there is a 4. This means the 2 will stay the same. 248 becomes 200 when rounded to the nearest hundred.

Rounding to the nearest ten or hundred can sometimes make addition or subtraction problems easier.

For example, use rounding to estimate the sum of 49 and 253. If 49 is rounded to the nearest ten, it becomes 50. If 253 is rounded to the nearest ten, it becomes 250. Now, add 50 and 250: $50 + 250 = 300$. In this case, rounding made the addition much simpler and is still close to the actual answer of 302.

Place Value Strategy

One- and two-step **addition and subtraction problems** can be solved using place value. For example, 432 + 826 can be thought of as a group of 4 hundreds, 3 tens, and 2 ones, plus 8 hundreds, 2 tens, and 6 ones. Combine the place values to solve:

$$4 \text{ hundreds} + 8 \text{ hundreds} = 12 \text{ hundreds}$$

$$3 \text{ tens} + 2 \text{ tens} = 5 \text{ tens}$$

$$2 \text{ ones} + 6 \text{ ones} = 8 \text{ ones}$$

$$1{,}200 + 50 + 8 = 1{,}258$$

Two-Step Problems

Consider the scenario: A shopper starts with \$456 in their wallet and then spends \$128 on shoes. Later that day, the shopper finds a 20-dollar bill in their pocket. Altogether, how much money does the shopper have left after shoe shopping?

The scenario can be solved in two steps. The shopper starts with \$456 and then spends \$128. This means that \$128 have been subtracted from \$456. \$456 consists of 4 hundreds, 5 tens, and 6 ones. \$128 consists of 1 hundred, 2 tens, and 8 ones. 8 ones cannot be subtracted from 6 ones, which means **regrouping** needs to occur.

4 hundreds, 5 tens, and 6 ones becomes 4 hundreds, 4 tens, and 16 ones. Now subtraction can continue: $456 - 128 = 328$.

$$\begin{array}{r} 4\,5\,6 \\ -\,1\,2\,8 \\ \hline \end{array} \quad \rightarrow \quad \begin{array}{r} {\scriptstyle 4\;16} \\ 4\,\cancel{5}\,\cancel{6} \\ -\,1\,2\,8 \\ \hline 3\,2\,8 \end{array}$$

The shopper then finds \$20 in their pocket. This indicates adding 20: $328 + 20 = 348$.

The shopper has \$348 after shoe shopping.

Example

Find the value of the expression below:

$$(648 + 250) - 30$$

648 consists of 6 hundreds, 4 tens, and 8 ones. 250 consists of 2 hundreds, 5 tens, and 0 ones. Combine each place value to find the sum.

$$600 + 200 + 40 + 50 + 8 + 0 = 898$$

Now, subtract 30, which consists of 3 tens, from 898. 898 consists of 9 tens. Subtract 3 tens from this: $898 - 30 = 868$.

$$(648 + 250) - 30 = 868$$

The answer is 868.

EXAMPLE

Find the value of the expression below:

$$640 - 180$$

Think about 640 and 180 in terms of place value.

$$640 = 6 \text{ hundreds} + 4 \text{ tens} + 0 \text{ ones}$$

$$180 = 1 \text{ hundred} + 8 \text{ tens} + 0 \text{ ones}$$

640 has 4 tens, which is not enough to subtract 8 tens from.

Regrouping is necessary. 1 of the hundreds becomes 10 tens. So, 6 hundreds + 4 tens + 0 ones becomes 5 hundreds + 14 tens + 0 ones.

Now subtraction can begin.

(5 hundreds + 14 tens + 0 ones) − (1 hundred + 8 tens + 0 ones) becomes

4 hundreds + 6 tens + 0 ones. In standard form, this is 460.

$$640 - 180 = 460$$

The answer is 460.

PICTURES AND MODELS

Use base ten blocks to represent 185 taken from 2 groups of 450 to get 715.

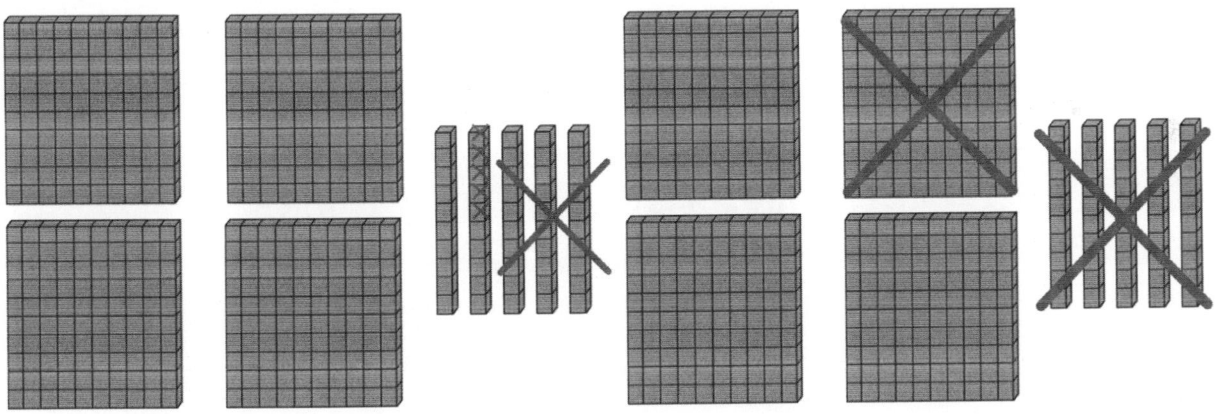

Number Lines

Make two jumps of 450, then back 185.

$$(450 + 450) - 185 = 715$$

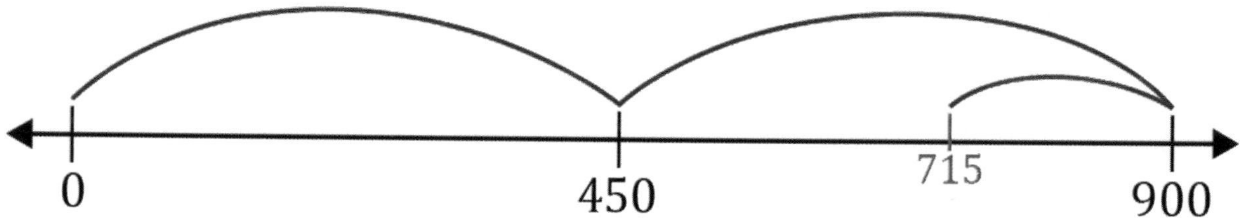

Equations

$$m = (450 + 450) - 185$$

The equation simplifies to $m = 715$, or \$715.

Example

Use the number line to represent $200 + 50 - 125$.

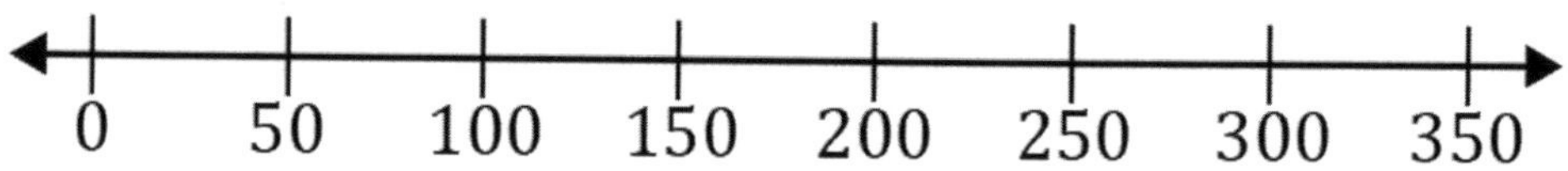

Start by adding 50 to 200. Make one jump forward from 200 to 250.

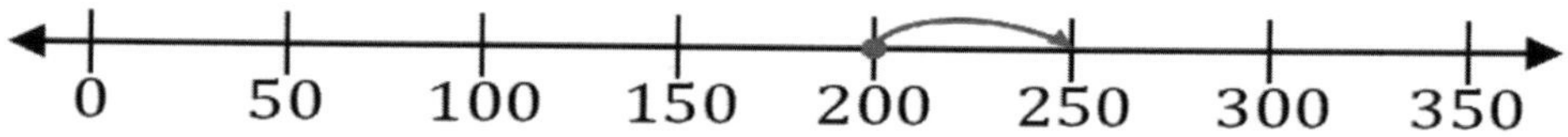

Now subtract 125. Jump back 100 and then back 25 more. 250 becomes 150 and then 125.

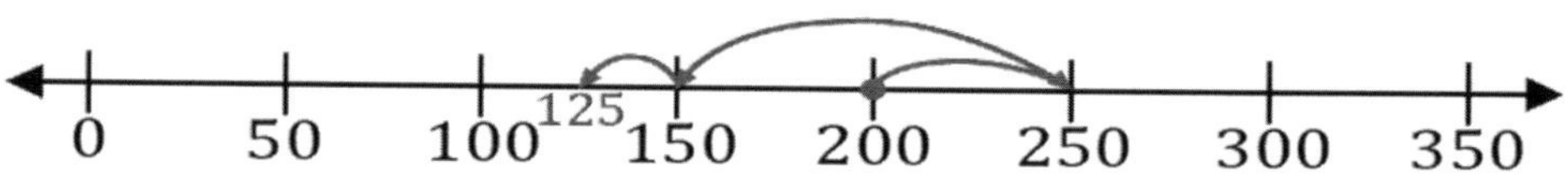

$$200 + 50 - 125 = 125.$$

Representing Multiplication

Repeated Addition

Multiplication is the process of counting equally sized groups. 5 groups of 7 means $7 + 7 + 7 + 7 + 7$. 5 groups of 7 is 35. The process of using addition to solve multiplication problems is known as **repeated addition**. This strategy is a great choice in many scenarios.

Number Lines

A number line can be used to show repeated addition. The **number line** below shows 4 equal hops of 3. 4 hops of 3 represents 4×3, which is 12: $3 + 3 + 3 + 3 = 12$.

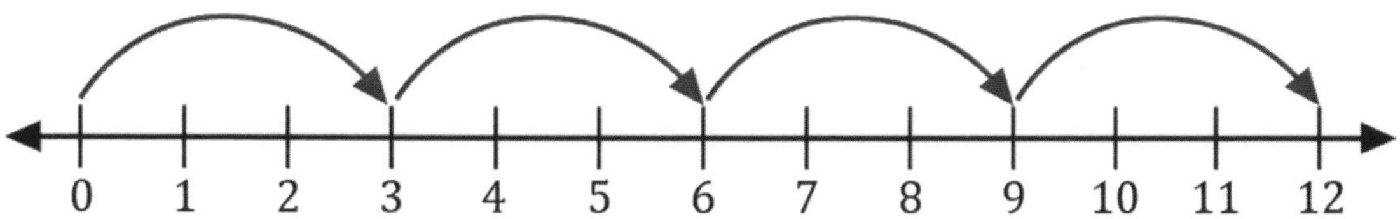

Arrays

An array is a collection and rows and columns representing a multiplication problem. This **array** has 10 columns and 4 rows. This array represents $10 \times 4 = 40$.

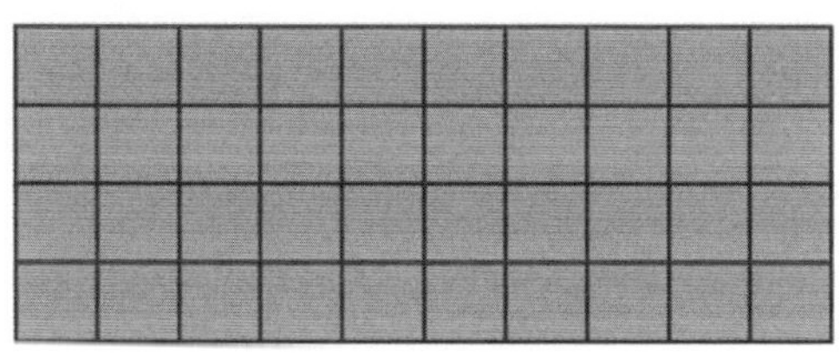

2 rows of 2 columns: $2 \times 2 = 4$

3 rows of 5 columns: $3 \times 5 = 15$

4 rows of 4 columns: $4 \times 4 = 16$

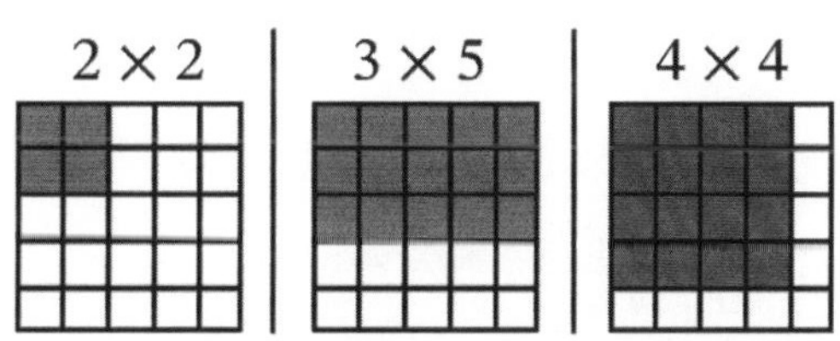

An **array** can be useful when solving real-world multiplication problems. For example, if a total of 7 blue fish are swimming in 3 separate groups, how many fish are there total?

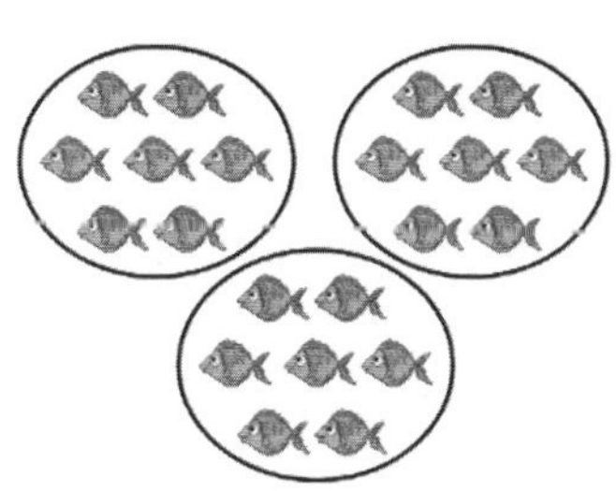

Drawing 3 groups of 7 is similar to creating an array of 3×7. Instead of creating 3 rows of 7 columns, the fish can be arranged into groups with 7 fish in each group. The total number of fish can be added up to find that $3 \times 7 = 21$.

Recalling Multiplication Facts

It is a great accomplishment to memorize the multiplication and division facts from 0 to 10. This is a goal that can take some time, so keep practicing!

Some strategies that can help:

1. **Skip count by the larger number.** For example, for the problem 8×3 skip count by 8s not 3s. This can save some time. $8 + 8 + 8$ is quicker than $3 + 3 + 3 + 3 + 3 + 3 + 3 + 3$.
2. **Use facts that you do know to help you.** For example, if you are stuck on 7×4 but you know 5×7 is 35, use logic to help you. 5 groups of 7 is 7 more than 4 groups of 7, so subtract 7 from 35 to solve 4×7.
3. **Use multiplication to help with division.** Multiplication and division are inverse operations. This means they undo each other. For example, $9 \times 4 = 36$ and $4 \times 9 = 36$ represent the inverse of $36 \div 4 = 9$ and $36 \div 9 = 4$. Your multiplication facts and division facts go hand-in-hand! Remember, multiplication and division fact families use the same three numbers. For example, $8 \times 7 = 56$, $7 \times 8 = 56$, $56 \div 7 = 8$, and $56 \div 8 = 7$ all use the same three numbers.

Strategies for Multiplication

Strategies such as mental math, skip counting, and drawing a model are great for single digit multiplication problems. However, for larger numbers, these strategies are less ideal. For example, solving 7×82 by skip counting would take quite a long time. 7, 14, 21, 28... all the way up to 574! Luckily, there are faster strategies for multiplying by two-digit numbers.

Standard Algorithm

Place the two-digit number over the one-digit number. Start by multiplying the numbers in the ones place: $2 \times 7 = 14$. Regroup the 14 into 1 and 4, as shown.

Now multiply 7 by the number in the tens place: $7 \times 8 = 56$. Add the 1 from up above: $56 + 1 = 57$. Write 57 down below: $82 \times 7 = 574$.

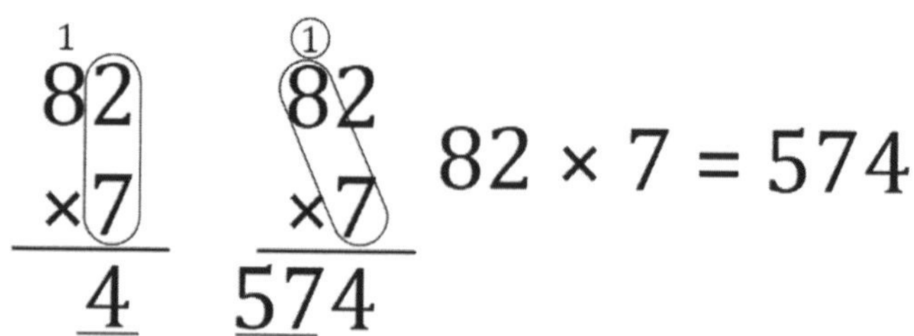

Partial Products

Think about the two-digit number in terms of its place value.

82 consists of 8 tens and 2 ones: $82 = 80 + 2$.

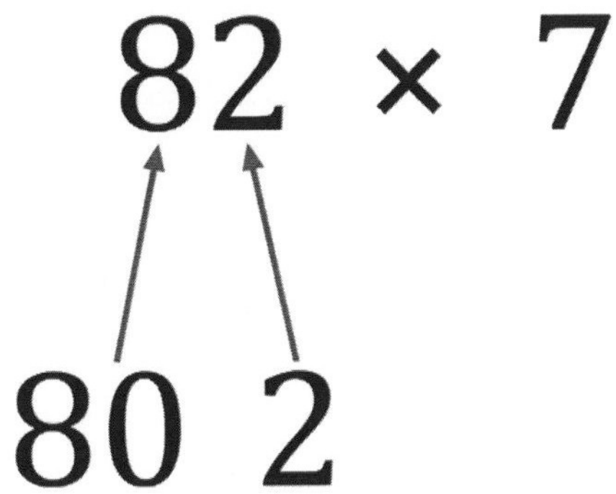

Now multiply each part by 7.

$$80 \times 7 = 560$$

$$2 \times 7 = 14$$

Now add the products.

$$\begin{array}{r} 5\ 6\ 0 \\ +\quad 1\ 4 \\ \hline 5\ 7\ 4 \end{array}$$

$$560 + 14 = 574$$

Distributive Property

Using the **distribute property** is very similar to using **partial products**.

Distribute the 7 to each part of 82. 7×80 and 7×2.

$$7 \times 82$$

$$(7 \times 80) \; + \; (7 \times 2)$$

$$(7 \times 80) + (7 \times 2) = 560 + 14 = 574$$

Commutative Property

The **order** of the **factors** does not matter in multiplication.

$$3 \times 9 \times 4 = 4 \times 9 \times 3$$

Associative Property

In multiplication, the **arrangement** of **grouping symbols** does not affect the outcome.

$$(4 \times 9) \times 3 = 4 \times (9 \times 3)$$

Equal Groups

When a set of numbers are partitioned, or divided, into equal pieces, multiplication and division will be involved. For example, consider the following scenario:

There are 28 kids at a bowling party, and there are 7 lanes available. How many bowlers will be in each lane?

This kind of problem can be solved using a variety of strategies.

Solve Using a Picture or Model

Draw a quick sketch of 7 bowling lanes. There are 28 bowlers, so fill each lane one by one. Each lane holds 4 bowlers.

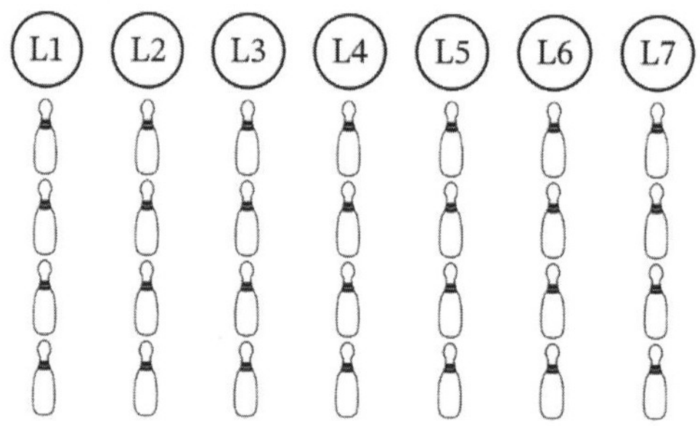

Solve Using Multiplication

7 groups of what equals 28? In other words, $7 \times ? = 28$.

$$7 \times 1 = 7$$

$$7 \times 2 = 14$$

$$7 \times 3 = 21$$

$$7 \times 4 = 28$$

Each lane will hold 4 bowlers.

Solve Using Division

$$28 \div 7 = ?$$

28 separated into 7 equal groups leaves 4 in each group.

$$28 \div 7 = 4$$

Even and Odd Numbers

You can determine if a number is even or odd by looking at the **ones** digit. Remember, the ones digit is the farthest digit to the right. If the ones digit can be divided, or is divisible, by 2, then the entire number is **even**. If the ones digit is not divisible by 2, then the entire number is **odd**. A number is divisible by 2 if it can be split into two groups without any remainders. For example, consider the number 3,916. The number 6 is in the ones place, and 6 is divisible by 2. Since the ones digit is divisible by 2, the number 3,916 is even. However, a number such as 83,097 is odd because the ones digit is 7, and 7 is not divisible by 2.

Numbers Divisible by 2

One way to determine if a number is divisible by 2 is to create a quick sketch or model. For example, consider the number 845,029. In this case, the ones digit is 9. Create a sketch with two groups and see if 9 can be divided into two groups with no remainders.

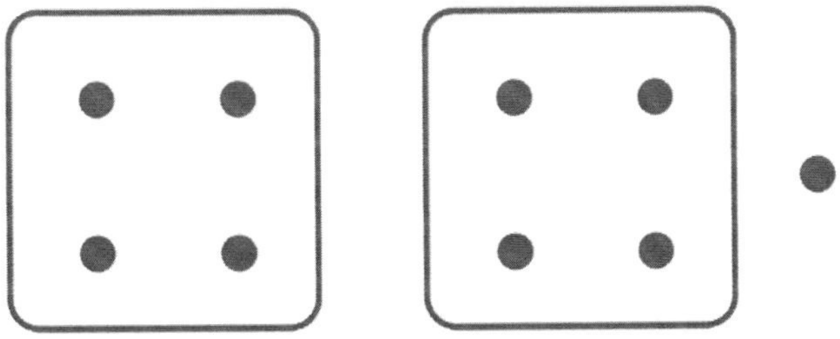

Notice that 9 cannot be split into two equal groups. It is close, but there is a remainder of 1. This means the number 845,029 is odd.

Now, consider the number 79,208. The ones digit is 8. The number 8 can be split into two equal groups, so 8 is divisible by 2. Therefore, 79,208 is even.

Quotients

Multiplication can be used to determine a **quotient**. Similar to addition and subtraction, multiplication and division are **inverse operations**. For example, $72 \div 9$ can be solved using the multiplication facts for 9. $72 \div 9 =$? Ask the question, "9 groups of what equals 72?"

$9 \times 6 = 54$ ... $9 \times 7 = 63$ ... $9 \times 8 = 72$. The missing number is 8. This means $72 \div 9 = 8$. This strategy is quite helpful because once the multiplication facts are automatic, division facts become automatic as well.

Consider the following examples:

$42 \div 7 =$? "7 groups of what equals 42?" $6 \times 7 = 42$. The product of 6 and 7 is 42. The quotient of 42 and 7 is 6.

$$42 \div 7 = 6$$

$48 \div 8 =$? "8 groups of what equals 48?" $6 \times 8 = 48$. The product of 6 and 8 is 48. The quotient of 48 and 8 is 6.

$$48 \div 8 = 6$$

Solving Multiplication and Division Problems

Multiplication and division can be **modeled** in many ways. Consider the problem 6×3. This can be solved using strategies such as **repeated addition**, **fact families**, **equal groups**, an **array**, **skip counting**, or a **tape diagram**. There are many options to choose from!

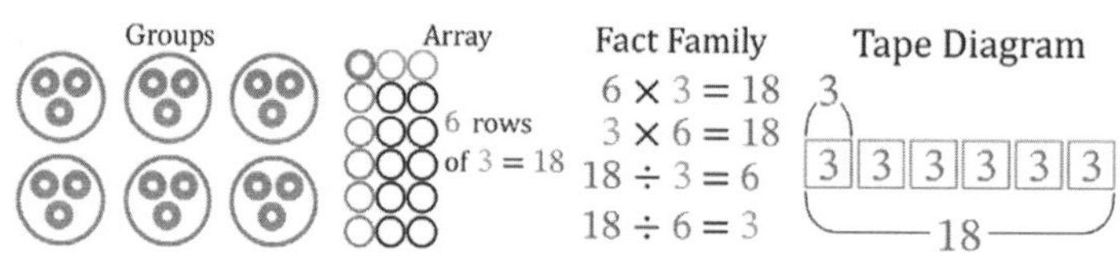

Skip Counting

3, 6, 9, 12, 15, 18

Repeated Addition

$3 + 3 + 3 + 3 + 3 + 3 = 18$

Two-step problems can be solved using the same strategies. Over time, these strategies will be replaced by memory.

EXAMPLE

Write a multiplication sentence that matches the tape diagram below.

The model shows 3 groups of 4. This can be expressed as the multiplication sentence $3 \times 4 = 12$.

EXAMPLE

Draw a model to represent $(15 \times 3) \div 5$.

The model shows 3 groups of 15, which equals 45.

45 is then divided into 5 equal groups. Each group contains 9 units.

$$(15 \times 3) \div 5 = 9$$

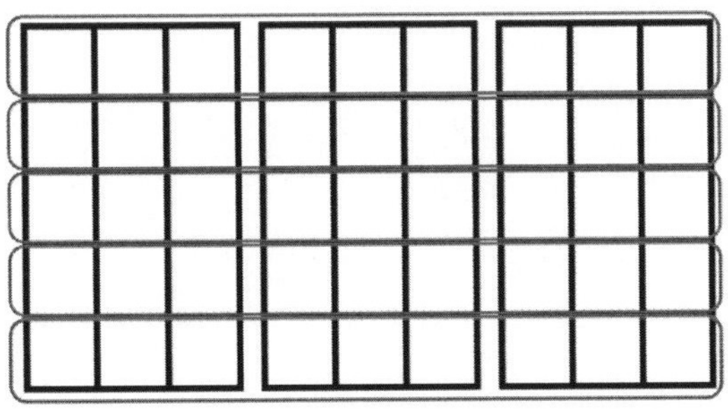

ONE-STEP PROBLEMS

A rancher has 12 acres of farmland, and 8 sheep live on each acre. How many sheep does the rancher have in all?

This problem can be solved in one step. Think about it as a picture or a model. There are 12 acres, and each acre contains 8 sheep. 12 groups of 8 means 12×8. To find 12×8, break it down into $(10 \times 8) + (2 \times 8)$ and then $(80) + (16) = 96$. The rancher has 96 sheep.

Now, try a two-step problem. This time, consider using a picture or a model to solve.

TWO-STEP PROBLEMS

A florist is creating flower arrangements for an event. 72 bouquets can be arranged in 3 hours. How many bouquets can be arranged in 15 minutes?

72 bouquets divided into 3 hours means 24 bouquets in each hour.

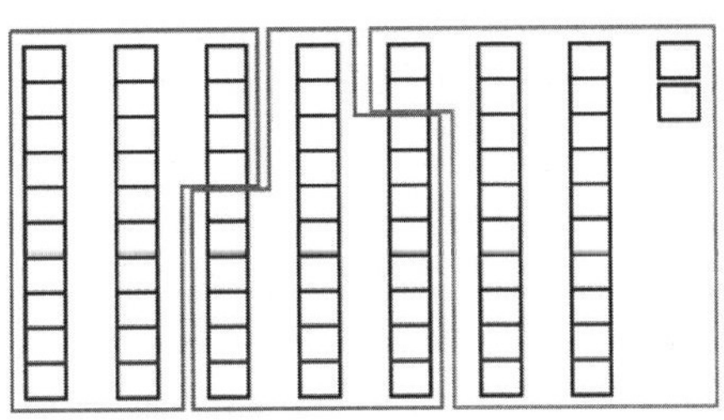

15 minutes is $\frac{1}{4}$ of an hour. We can see this by looking at a clock. The minute hand of a clock at 12:15 is $\frac{1}{4}$ through the circle.

So, divide 24 by 4.

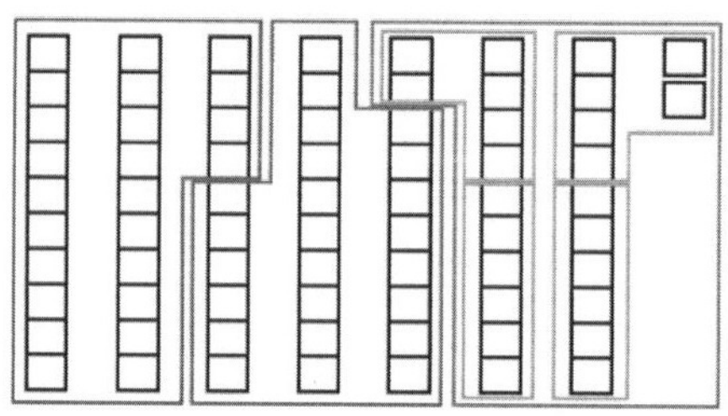

$$24 \div 4 = 6$$

So, 6 bouquets can be made every 15 minutes.

Fractions

Unit Fractions

A **unit fraction** is any fraction with a numerator of 1. For example, $\frac{1}{2}, \frac{1}{3}, \frac{1}{4}$, and $\frac{1}{5}$ are considered unit fractions. When a unit fraction is added to itself multiple times, the result is a larger fraction. For example, $\frac{1}{8} + \frac{1}{8} + \frac{1}{8} = \frac{3}{8}$.

The opposite of this is **decomposing** a fraction into the sum of its unit fractions.

$\frac{5}{6}$ is the result of combining $\frac{1}{6}$ five times.

$$\frac{5}{6} = \frac{1}{6} + \frac{1}{6} + \frac{1}{6} + \frac{1}{6} + \frac{1}{6}$$

Decomposing Fractions

When a fraction is decomposed, it is **split up** into **smaller** fractions. The sum of these smaller fractions is the original fraction. For example, $\frac{7}{8}$ can be split up into the sum of $\frac{1}{8} + \frac{1}{8} + \frac{1}{8} + \frac{1}{8} + \frac{1}{8} + \frac{1}{8} + \frac{1}{8}$. However, this is not the only way to decompose $\frac{7}{8}$.

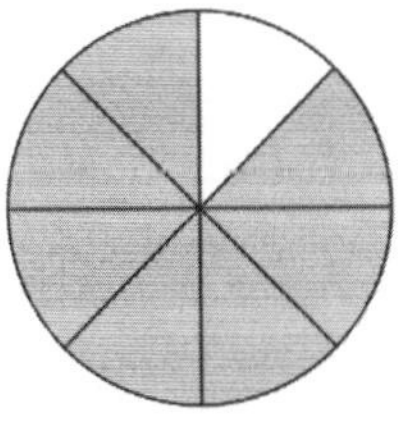

For example, the model below shows 7 shaded shares out of 8, or the fraction $\frac{7}{8}$. The fraction $\frac{7}{8}$ can be created in many ways. For example, $\frac{2}{8} + \frac{4}{8} + \frac{1}{8}$ has a sum of $\frac{7}{8}$.

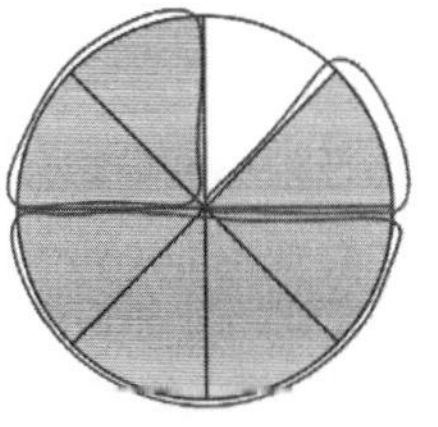

Adding and Subtracting Fractions with Like Denominators

When adding or subtracting fractions, it is important to make sure that the denominators are the same. If the fractions have like denominators, add the numerators together and leave the denominator the same. Consider the diagram below. Each whole is divided into 8 equal shares. The circle on the left shows 2 out of 8, or $\frac{2}{8}$. The circle on the right shows 6 out of 8, or $\frac{6}{8}$.

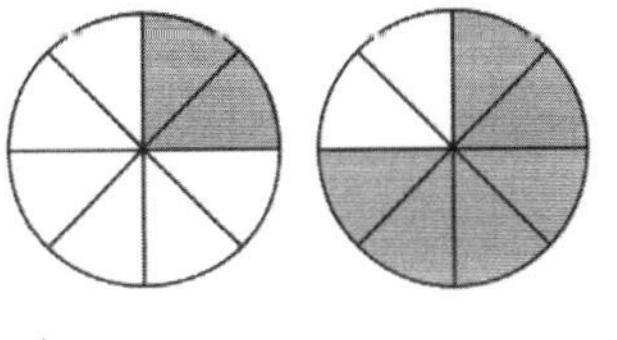

Both fractions have the same denominators so their numerators can be combined. $\frac{2}{8}+\frac{6}{8}$ combines to $\frac{8}{8}$. Another way to express $\frac{8}{8}$ is 1 whole.

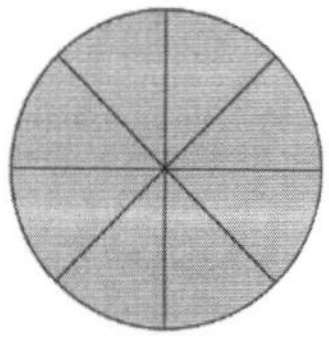

A picture or model is a great way to represent addition and subtraction of fractions. However, there is a shortcut! When denominators are the same, simply add or subtract the numerators.

$$\frac{2}{7}+\frac{4}{7}=\frac{6}{7}$$

Notice how the denominator, 7, stays the same and the numerators are added.

$$\frac{8}{9}-\frac{3}{9}=\frac{5}{9}$$

Notice how the denominator, 9, stays the same and the numerators are subtracted.

Example

Write and solve an addition equation that represents adding the top fraction to the bottom fraction illustrated below.

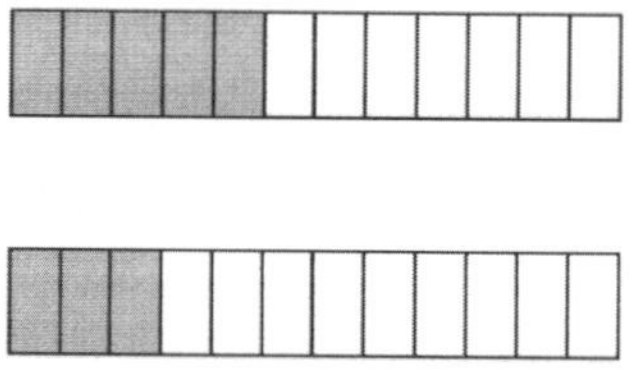

Top fraction: $\frac{5}{12}$

Bottom fraction: $\frac{3}{12}$

Both fractions have the same denominator, so they can be combined.

$$\frac{5}{12}+\frac{3}{12}=\frac{8}{12}$$

$\frac{8}{12}$ can be simplified by dividing the numerator and denominator by 4. $\frac{8}{12}$ becomes $\frac{2}{3}$.

$$\frac{5}{12}+\frac{3}{12}=\frac{8}{12}=\frac{2}{3}$$

EXAMPLE

Model the subtraction equation on the number line, then solve: $\frac{6}{10} - \frac{2}{10} = ?$

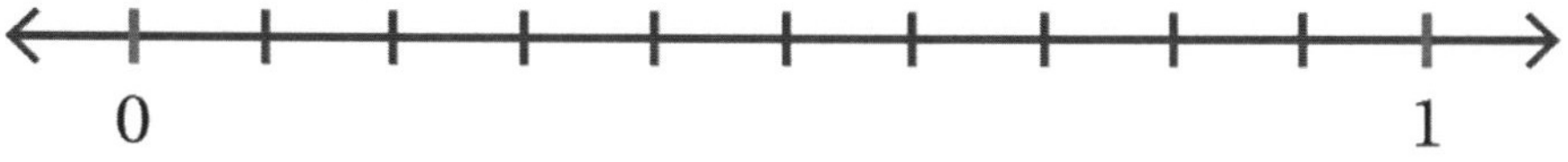

The number line is divided into 10 equal shares. Each hash mark represents $\frac{1}{10}$.

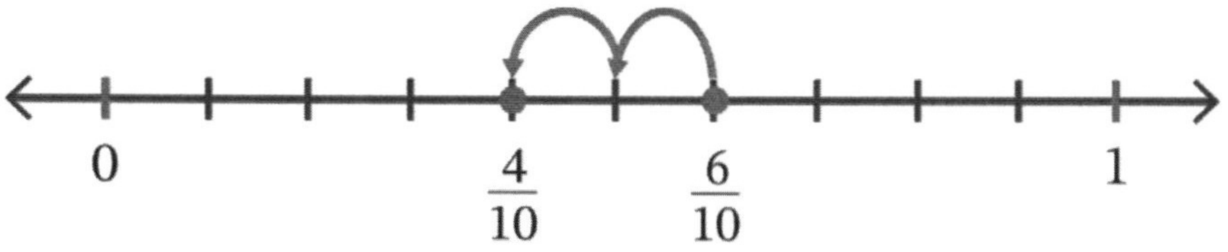

The red dot starts at $\frac{6}{10}$. The arrows show 2 jumps back, which represents subtraction of $\frac{2}{10}$: $\frac{6}{10} - \frac{2}{10} = \frac{4}{10}$.

Representing Real-World Relationships

Input-output tables are used to show the relationship between two variables. For example, consider the muffin pan below.

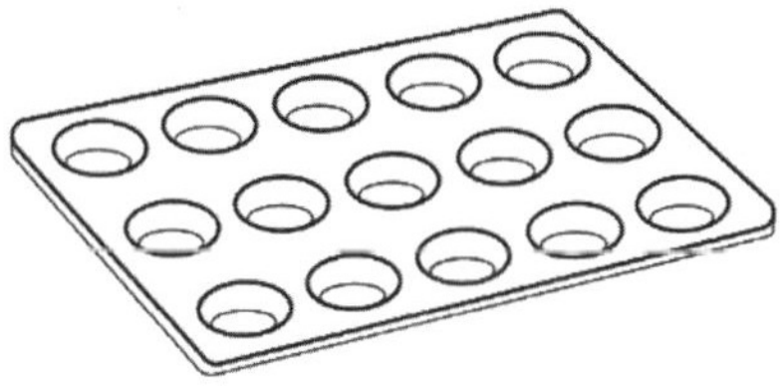

Each pan can hold 15 muffins. This means that 15 muffins can be baked in 1 pan, 30 muffins can be baked in 2 pans, 45 muffins can be baked in 3 pans, and so on. These values can be placed in an input-output table to show the relationship between the number of pans and the total number of muffins.

Pans	1	2	3	4
Muffins	15	30	45	60

The rule for this input-output table is number of muffins = number of pans × 15.

Here are some more examples:

Rule: Find y by adding 2 to x.

x	y
8	10
4	6
2	4
6	8
3	5

Rule: Find y by multiplying x by 4.

x	9	4	5	2
y	36	16	20	8

Example

Complete the table so that the rule applies. Rule: James earns $42 per hour

Hours	1	2	3	4	5	6	7
Dollars		84				252	

The rule states that $42 is earned each hour. The number of dollars can be determined by multiplying the number of hours by 42, or by adding 42 to the previous dollar amount.

$$1 \times 42 = 42$$

$$2 \times 42 = 84$$

$$3 \times 42 = 126$$

$$4 \times 42 = 168$$

$$5 \times 42 = 210$$

$$6 \times 42 = 252$$

$$7 \times 42 = 294$$

Hours	1	2	3	4	5	6	7
Dollars	42	84	126	168	210	252	294

EXAMPLE

Write a rule that describes the relationship in the table.

x	4	5	6	7
y	32	40	48	56

Each value of y is 8 times as large as the corresponding x-value: $y = 8 \times x$. This can be expressed as "y is 8 times the value of x" or "multiply x by 8 to determine the value of y."

Measurement and Geometry

Coins and Bills

The value of a collection of coins and bills can be found by adding the values according to place value.

Place Value

	Penny	$0.01 or 1¢
	Nickel	$0.05 or 5¢
	Dime	$0.10 or 10¢
	Quarter	$0.25 or 25¢
	1-dollar bill	$1.00
	5-dollar bill	$5.00
	10-dollar bill	$10.00
	20-dollar bill	$20.00

Adding Coins and Bills

When adding coins and bills, make sure to line up the decimals.

Example

Allison has one $1-dollar bill, two quarters, and 4 dimes in her bank. Find the total amount of money in Allison's bank.

It may help to draw a picture. Write the value of each bill or coin in its picture. Then add up the bills and coins to find the total money.

1 + 0.25 + 0.25 + 0.10 + 0.10 + 0.10 + 0.10 = 1.90

There is $1.90 in Allison's bank.

EXAMPLE

Start by combining the bills.

$5 + 1 + 1 + 1 = 8$. So far, there are $8.

Now add the coins. There are 2 quarters, 1 nickel, and 2 dimes.

Be sure to line up the decimals when adding.

$0.25 + 0.25 + 0.1 + 0.1 + 0.05 = 0.75$, or 75 cents.

$$\begin{array}{r} 0.25 \\ 0.25 \\ 0.10 \\ 0.10 \\ +\,0.05 \\ \hline 0.75 \end{array}$$

Combine the bills and the coins: $8 + 0.75 = 8.75$, or $8.75.

$$\begin{array}{r} 8.00 \\ +\,0.75 \\ \hline 8.75 \end{array}$$

Remember, when the sum of a collection of coins exceeds 99 cents, the amount can be expressed using bills. For example, 99 pennies is 99 cents, or $0.99, but 100 pennies is 100 cents, or 1 dollar bill.

Comparing Values

Example

Compare the two sets of coins. Which set has a greater value?

Set A:

Set B:

Set A has a value of \$0.70. Set B has a value of \$0.55. Therefore, the value of Set A is greater than the value of Set B.

Making Change

When making change, subtract the amount paid from the amount given. For example, if Dan gave the cashier a \$5 for items that produced a total cost of \$2.88, the cashier would need to give Dan his change in return. To calculate the amount needed to return to Dan, the cashier would subtract \$2.88 from \$5.00. This represents the equation 5.00 - 2.88 = 2.12. Therefore, Dan will receive \$2.12 in change after paying for his items.

Measurement Systems

The **metric system** is a system of measurement used for identifying mass, volume, and length. Most countries in the world use the metric system when making calculations in math and science. However, the United States has another measurement system called the **customary system**. The customary system also measures mass, volume, and length, but it uses different units.

For example, the metric system has base units of meters, liters, and grams. Meters measure length, liters measure volume, and grams measure mass. The metric system is convenient in math and science because it follows the base-ten number system. This means that quick calculations and conversions can be done simply by multiplying or dividing by a power of ten. For example, 10 millimeters is equivalent to 1 centimeter.

The customary system has much more than three units. For example, when measuring volume, options include ounces, cups, pints, quarts, and gallons. Converting between units is more challenging because each unit has a different conversion factor. For example, there are 4 quarts in a gallon, but 16 ounces in a pint. Distance is measured in inches, feet, yards, and miles. Mass is similar to weight, and is measured in ounces, pounds, and tons. It is important to become familiar with units in the metric and customary systems.

Length of a Segment

Length can be measured by the following customary units: inches, feet, yards, miles.

- **1 inch**: about the width of a quarter
- **1 foot**: about the height of a 2 liter soda bottle (12 inches)
- **1 yard**: about the width of a door (3 feet)
- **1 mile**: about 4 laps on an outdoor running track

Length can also be measured with metric units. The base unit for length in the metric system is meters, but other measurements for length include: kilometer, hectometer, decameter, decimeter, centimeter, millimeter.

- **1 meter:** about the width of a standard door frame

Example

Find the length of the line segment using the ruler below. The ruler is in inches.

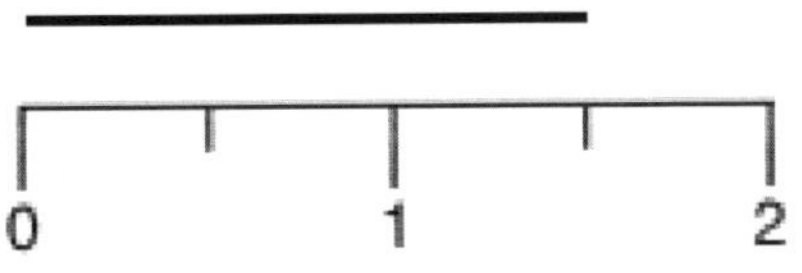

The line segment begins at 0, and ends at the mark that is halfway between 1 and 2 inches. Halfway between 1 and 2 inches is $1\frac{1}{2}$ inches. The line segment is $1\frac{1}{2}$ inches long.

Volume and Weight

Volume is the amount of space something fills. **Weight** is the heaviness of an object. The terms are similar, but not interchangeable. Volume is not the same as weight.

For example, consider the image below. The two spheres have the same volume because they take up the same amount of space. However, their weights are very different. One sphere weighs 12 pounds and the other weighs 1 pound. Remember, volume and weight do not describe the same thing. They are similar, but one describes how much space is filled by an object, and one describes the heaviness of an object.

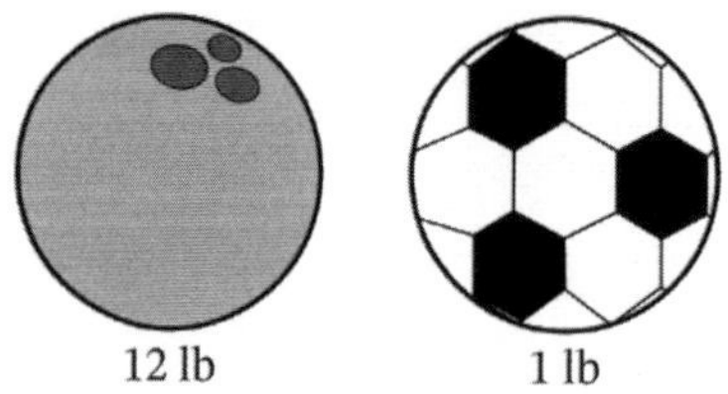

When determining how much space is filled, use volume. When determining the heaviness of an object, use weight.

Measuring Volume and Weight

Volume and weight are similar measurements but are not interchangeable. Volume measures capacity, or how much space is filled. Weight measures how heavy an object is.

Volume can be measured by the following customary units: ounces, cups, pints, quarts, gallons.

- **1 ounce**: the amount of liquid in two tablespoons
- **1 cup**: the amount of liquid in a coffee cup (8 ounces)
- **1 pint**: the amount of liquid in a small water bottle (16 ounces)
- **1 quart**: the amount of liquid in a jug of juice (2 pints)
- **1 gallon**: the amount of liquid in a large container of milk

Volume can also be measured with metric units. The base unit for volume in the metric system is liters, but other measurements for volume include: kiloliter, hectoliter, decaliter, deciliter, centiliter, milliliter.

- **1 liter:** about the volume of a large water bottle

Volume describes **capacity**. Capacity can be determined using tools such as a **beaker**. The amount of liquid in the beaker is 125 mL. This represents the volume.

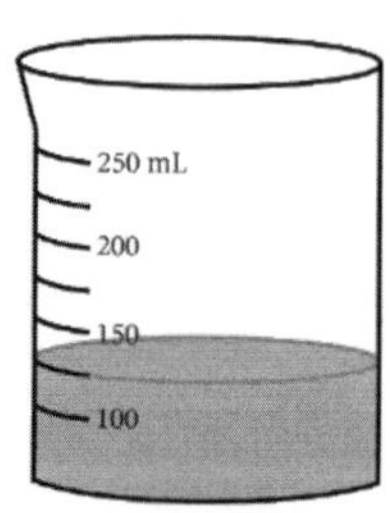

Weight can be measured by the following customary units: ounces, pounds, tons.

- **1 ounce**: about the weight of a candy bar
- **1 pound**: about the weight of a football
- **1 ton**: about the weight of a car

Weight can also be measured with metric units. The base unit for weight in the metric system is grams, but other measurements for weight include: kilogram, hectogram, decagram, decigram, centigram, milligram.

- **1 gram:** about the mass of a paperclip

Weight describes how **heavy** an object is. A **scale** can be used to determine weight. Weight is measured in ounces, pounds, or tons.

Classifying and Sorting Figures

Two-Dimensional Figures

Two-dimensional shapes can be classified by the number of sides they have. Polygons are considered **regular polygons** if all the sides are the same length. For example, a regular pentagon has 5 sides, and all of the sides are the same length. Side lengths that are the same are considered **congruent**. For example, a square has 4 congruent side lengths.

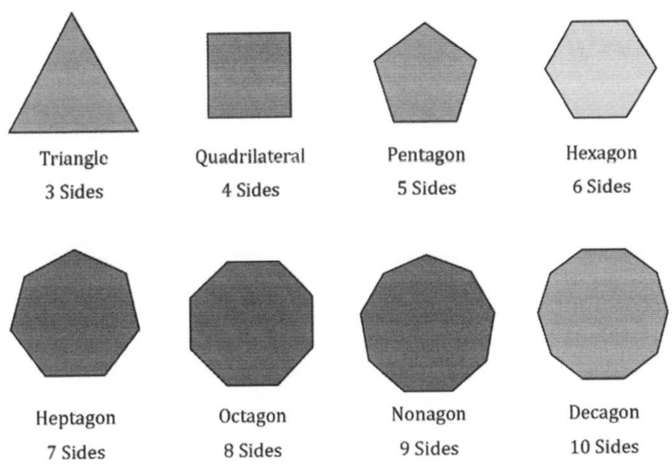

QUADRILATERALS

A **quadrilateral** is a 4-sided shape. There are many subcategories of quadrilaterals. The most common are parallelogram, rhombus, rectangle, square, and trapezoid.

Parallelogram: A parallelogram has two pairs of parallel sides. It sometimes looks like a leaning rectangle.

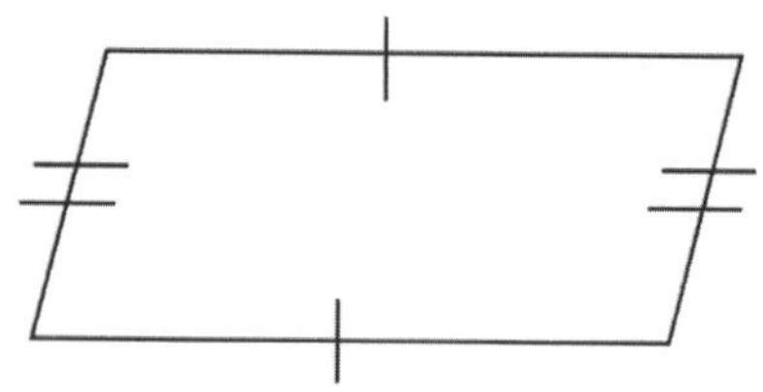

Rhombus: A rhombus is a parallelogram with four equal sides. It has no 90° angles.

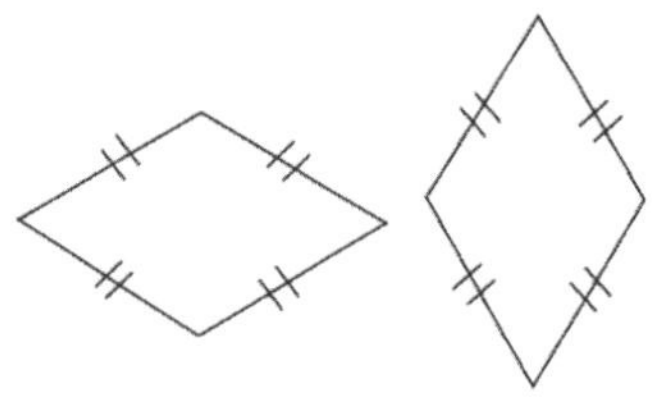

Rectangle: A rectangle has two equal parallel long sides, two equal parallel short sides, and four 90° angles.

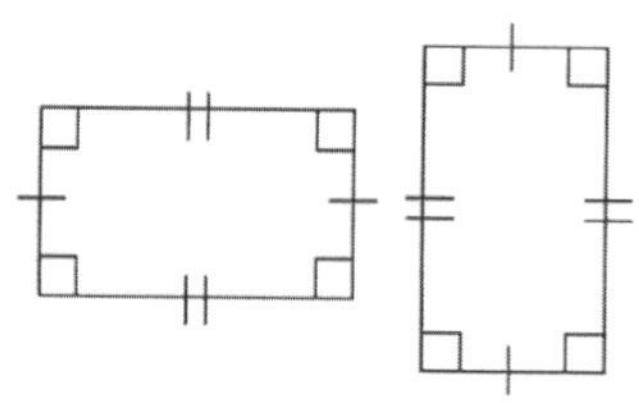

Square: A square has four equal sides. Its opposite sides are parallel, and it has four 90° angles.

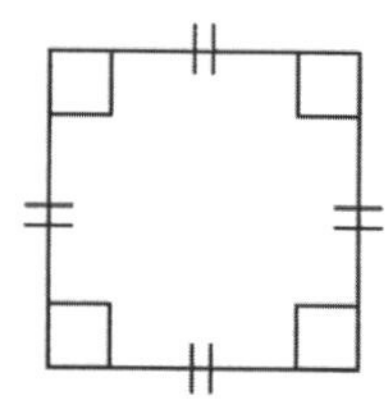

Trapezoid: A trapezoid has one pair of parallel sides and one pair of non-parallel sides.

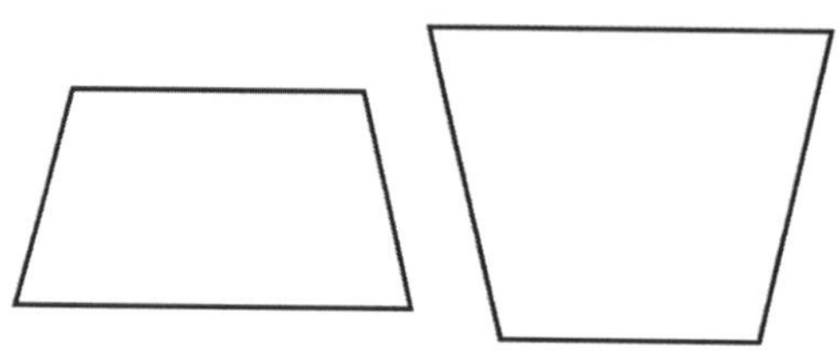

Not all 4-sided shapes fall into these categories. For example, the shapes below are quadrilaterals, but they do not fall into any of the subcategories mentioned above. They are simply referred to as quadrilaterals.

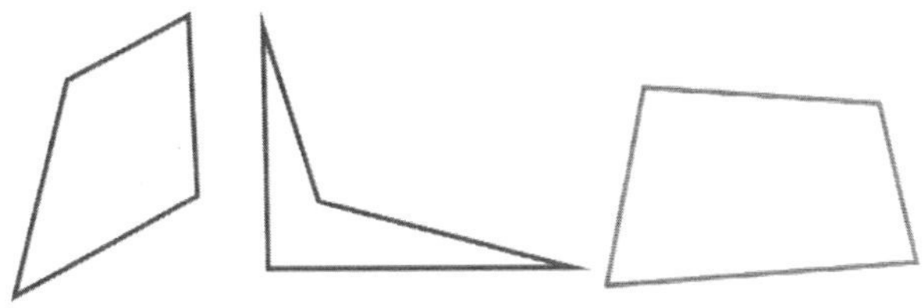

EXAMPLE

All the shapes below can be classified as ________. The top two shapes can also be classified as ________ or as ________.

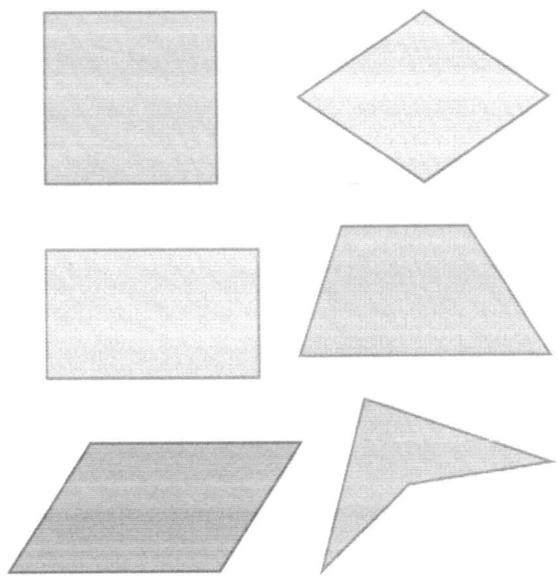

Quadrilateral: 4-sided polygon
Parallelogram: A quadrilateral that has 2 pairs of parallel sides
Rhombus: A parallelogram with 4 congruent side lengths

All the shapes below can be classified as quadrilaterals. The green and yellow shapes can also be classified as rhombi or as parallelograms.

Perimeter

Perimeter is the distance around the outside of a two-dimensional shape. Perimeter can be calculated by finding the **sum** of a shape's **side lengths**. When a shape has **congruent** side lengths, all the sides are the same length. These shapes are called **regular polygons**.

The shape below has 6 congruent side lengths. Find the sum of the side lengths to calculate the perimeter of the regular hexagon. One side length is labeled 3 inches, therefore all of the side lengths are 3 inches. Perimeter $= 3 + 3 + 3 + 3 + 3 + 3$, which simplifies to $P = 18$. The perimeter of the regular hexagon is 18 inches.

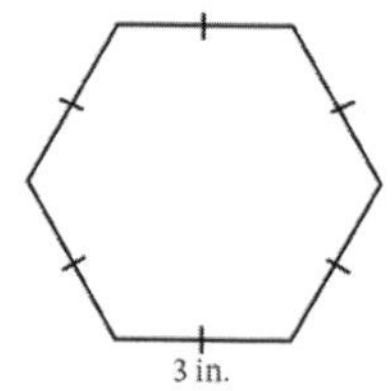

The same rules apply for polygons with missing side lengths. The triangle below has 3 congruent side lengths, and the perimeter is 18 cm. Ask the question, "$\_ + \_ + \_ = 18$?" Since $6 + 6 + 6 = 18$, the side lengths are all 6 cm.

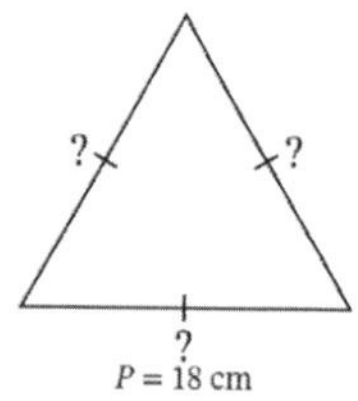

Example

A rectangle has a length of 3 inches and a width of 1 inch. Find the perimeter of the rectangle.

Using a ruler, draw a model of the rectangle with the given side lengths.

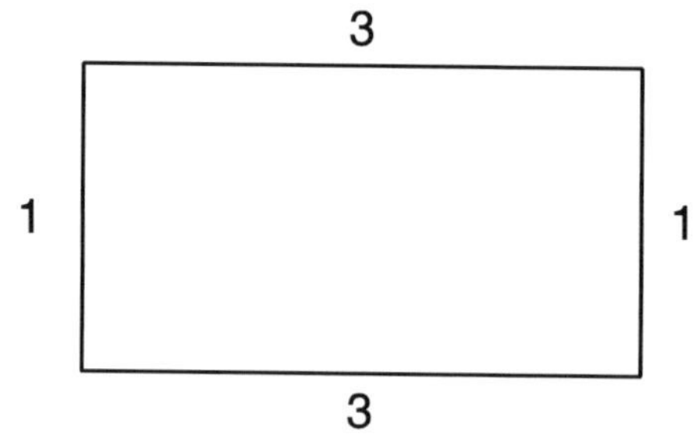

The perimeter of a rectangle is the sum of side lengths: perimeter = 3 inches + 1 inch + 3 inches + 1 inch = 8 inches.

Example

A triangle has sides that are 6 cm, 8 cm, and 11 cm long. Find the perimeter of the triangle.

Using a ruler, draw a model of the triangle with the given side lengths.

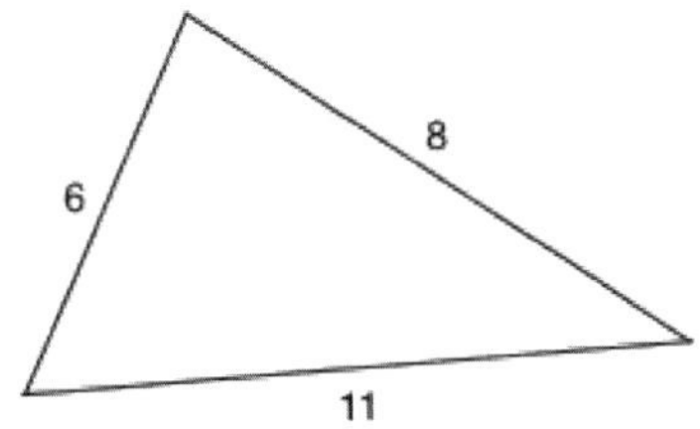

The perimeter is the sum of the side lengths: 6 cm + 8 cm + 11 cm = 25 cm.

EXAMPLE

A carpenter is building an octagonal window for a boat. The outer octagon has side lengths of 8 inches, and the inner octagon has side lengths of 7 inches. What is the difference in the perimeters of the two octagons?

Perimeter of the outer octagon: $8+8+8+8+8+8+8+8=64$. Therefore, the perimeter is 64 inches.

Perimeter of the inner octagon: $7+7+7+7+7+7+7+7=56$. Therefore, the perimeter is 56 inches.

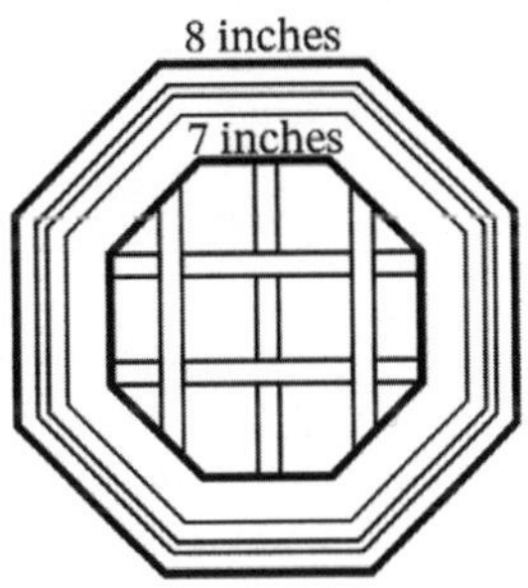

$64-56=8$, so the difference between the two perimeters is 8 inches.

Area

AREA OF A RECTANGLE

The **area formula** for a **rectangle** is Area $=$ Length $\times$ Width. This means the area can be determined by multiplying the number of **rows** by the number of **columns**. Think about the area of a rectangle as an array. For example, a rectangle that is 6 ft by 4 ft is essentially 6 columns and 4 rows. 6 groups of 4 is the same thing as 4 groups of 6, so the area can be calculated as 6×4 or 4×6. $6\times4=24$ and $4\times6=24$. The area is 24 square units.

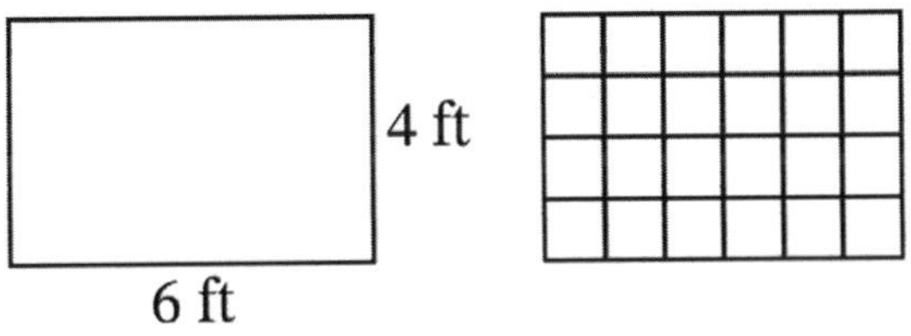

EXAMPLE

Draw a rectangle that is 8 units high and 10 units wide.

10 units

8 units

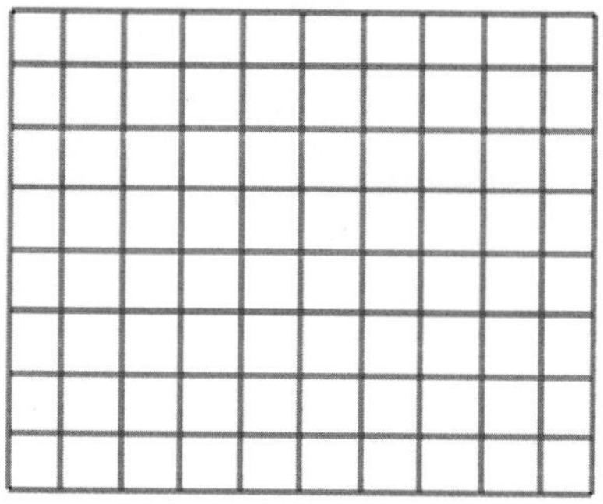

To find the area, count the squares inside the area of the rectangle. There are 80 squares; the area of the rectangle is 80 units^2.

EXAMPLE

Draw a square that is 7 units wide and 7 units high. Here, each circle represents one unit.

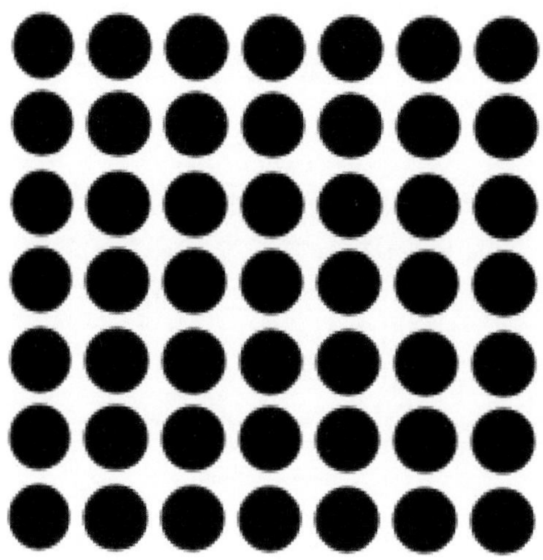

The total number of circles is the area. There are 49 circles, so the area of the square is 49 units^2.

EXAMPLE

A pan of fudge needs to be cut into equally sized pieces. One row is already divided. How many pieces of fudge will there be in the entire pan?

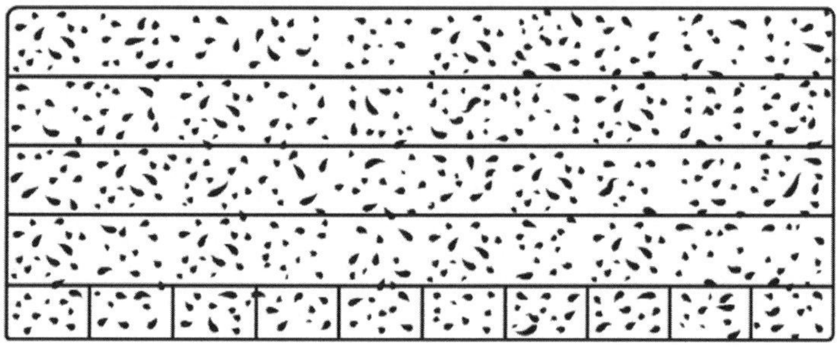

The total number of pieces can be calculated by multiplying the number of pieces in one row, by the total number of rows. This can be thought of as an array of 5 rows with 10 columns. 5 groups of 10 equals 50. The entire pan holds 50 pieces of fudge.

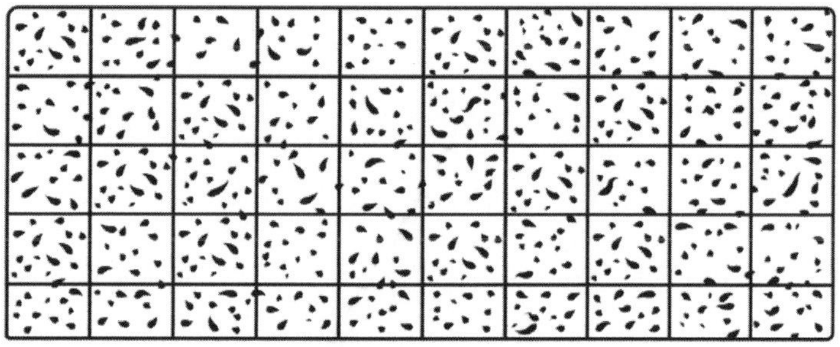

EXAMPLE

A rectangle has an area of 16 unit squares. The rectangle has two columns. How many rows does the rectangle have?

The area of the rectangle is 16 unit squares. The rectangle has 2 columns, so it must have 8 rows: $2 \times 8 = 16$. The rectangle forms an array of 2 columns and 8 rows.

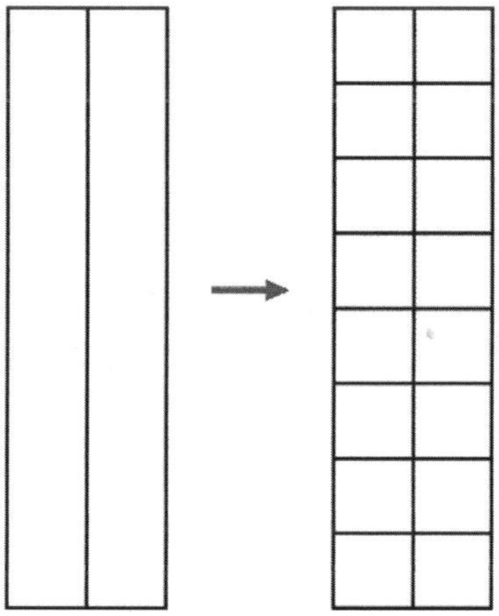

EQUAL PARTS AND AREA

There is more than one correct way to decompose, or divide, a two-dimensional figure. For example, if a pan of brownies needs to be sliced into 8 equal pieces, there are a few ways to achieve this. Consider the rectangular brownie pans below. Notice how both pans have 8 **equally sized** brownies, even though the identical pieces are not the same shape for both pans. Each brownie can be represented as the unit fraction $\frac{1}{8}$. Remember, a unit fraction has a numerator of 1.

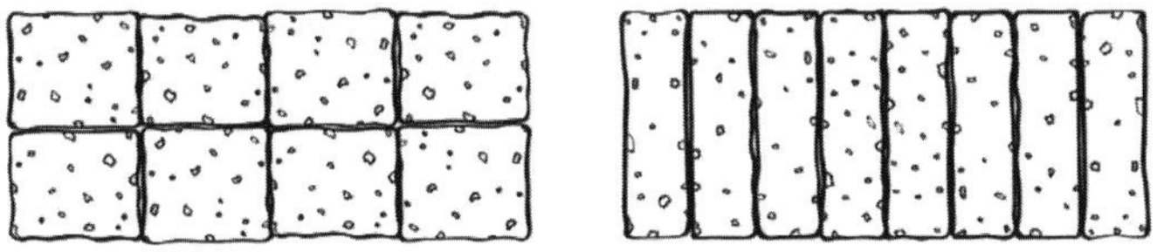

Congruency

Congruent figures are the same size and shape. This means that all of the sides are the same length, and all of the angles are the same measure. The two rectangles below are congruent.

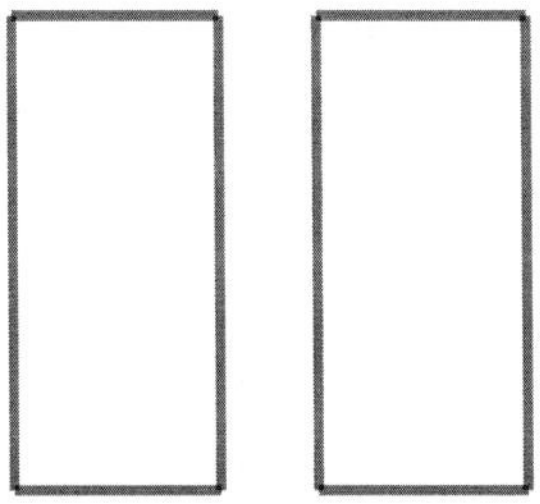

CONGRUENT PARTS OF A SQUARE

The congruent parts are the parts of the square that are the same size. On a square, all side lengths are equal and all angles are equal. The congruent parts are the side and angles. For example:

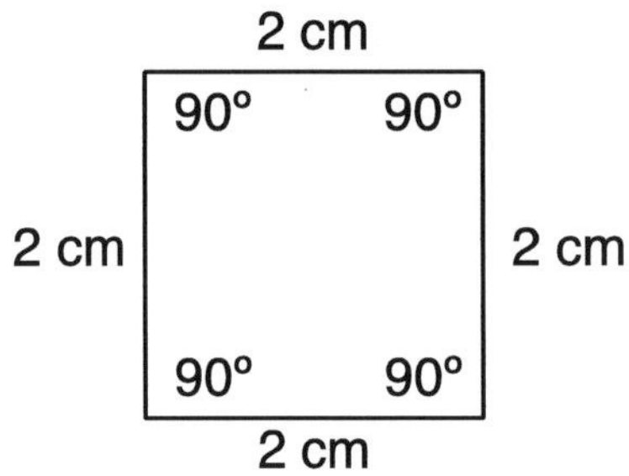

Congruent Parts of a Parallelogram

A parallelogram is a quadrilateral where pairs of opposite sides are parallel. Opposite sides of a parallelogram are congruent, and opposite angles are also congruent. For example:

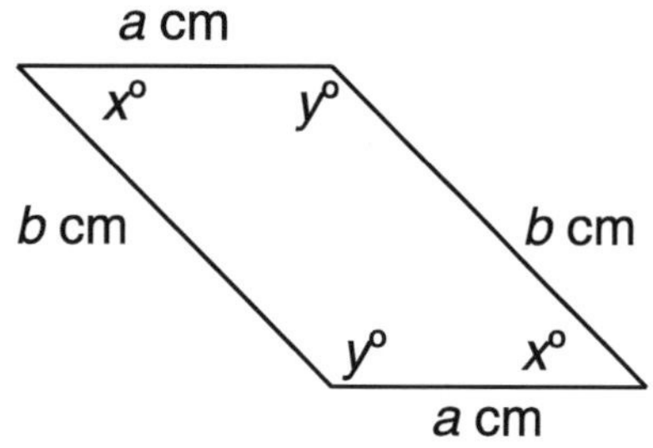

Congruent Triangle

Example

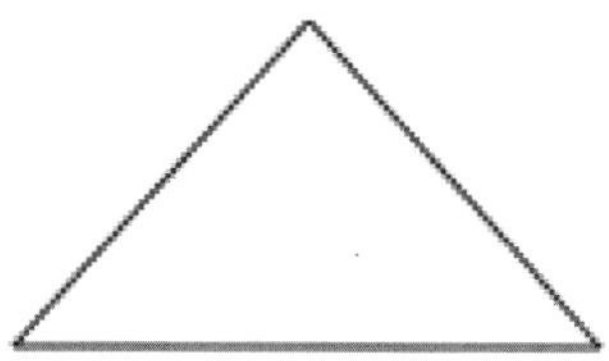

A congruent triangle can be made by moving the original triangle left, right, up, or down. The congruent figure below, with the dashed lines, was moved up and to the right.

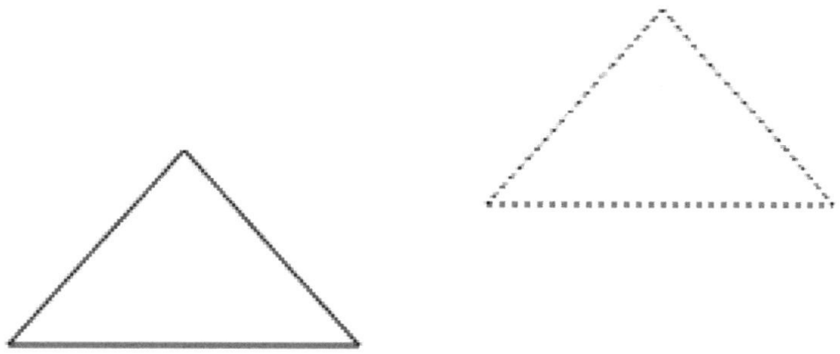

Example

Create a congruent triangle to the one below by turning the triangle.

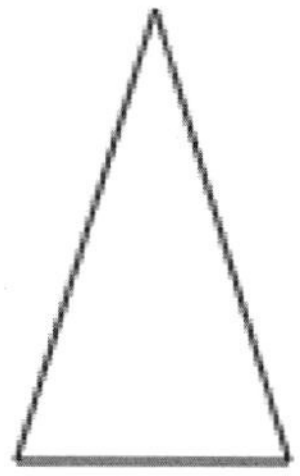

To create a congruent triangle by rotating, first draw a point of rotation. The point can be outside, inside, or on the figure. Then pick a number of degrees to rotate the figure. For example, rotate the figure 90º.

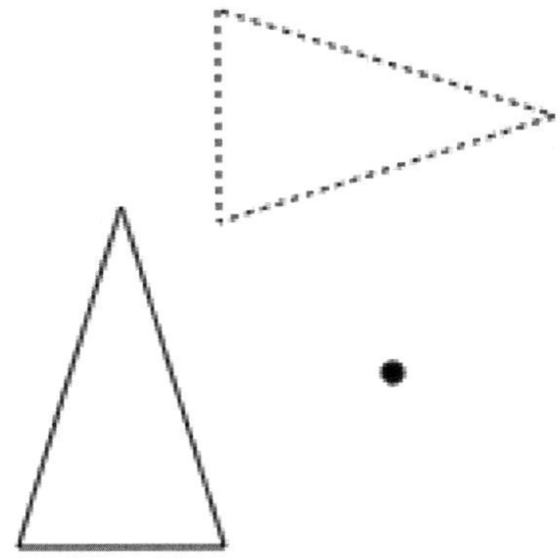

Points, Lines, Rays, and Angles

A **point** is an exact location with no size. This often appears as a dot, and is usually labeled with a letter, such as Point A.

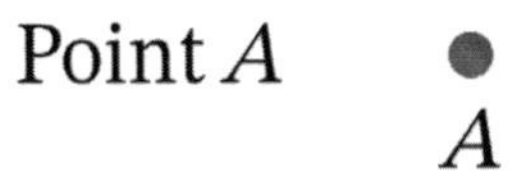

A **line** is a straight path that extends in opposite directions. A line does not have a width, and it extends indefinitely in both directions. Lines are labeled with letters representing points on the line, such as $\overleftrightarrow{AB}$.

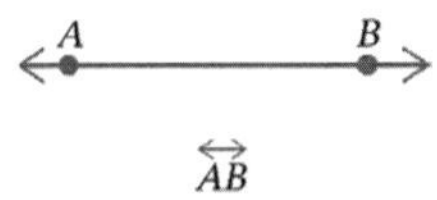

A **line segment** is a part of a line. A line segment has endpoints, beginning and ending at a point labeled with a letter, such as $\overline{AB}$.

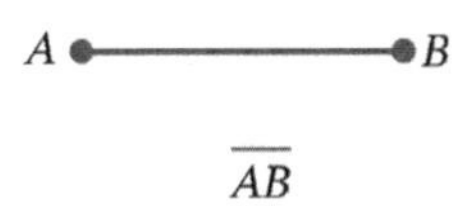

A **ray** is a part of a line that starts at an endpoint and then extends out indefinitely. Rays are labeled starting with a letter at the endpoint and ending with a letter representing another point on the ray, such as $\overrightarrow{AB}$.

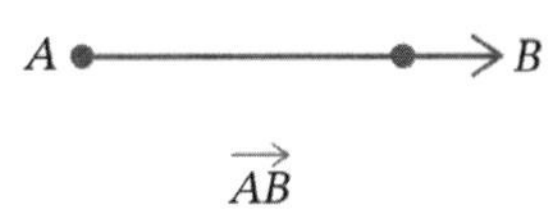

An **angle** is formed when two straight lines or rays meet at a common endpoint. The two rays below form an angle.

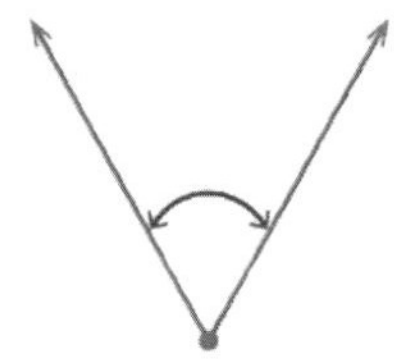

Parallel lines extend side-by-side and never intersect. To help visualize parallel lines, think of train tracks.

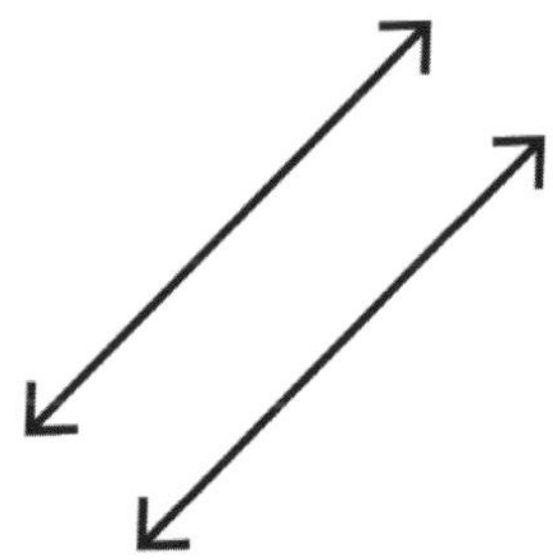

Perpendicular lines intersect to form a right angle. To help visualize perpendicular lines, think of a corner or a T-shape.

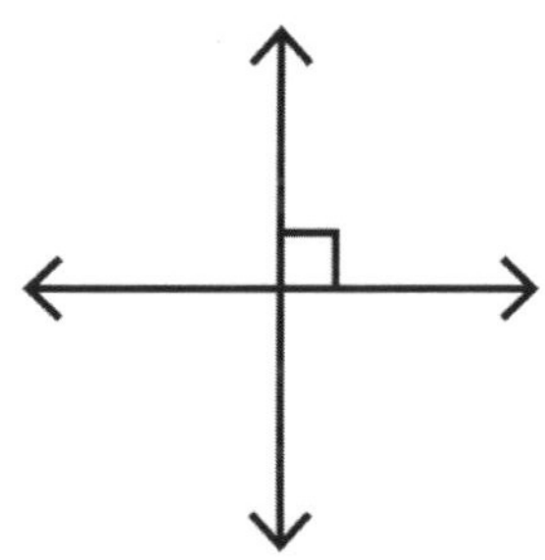

Review Video: Parallel and Perpendicular Lines
Visit mometrix.com/academy and enter code: 815923

Time

Draw the time 7:20 on an analog clock and a digital clock.

On the analog clock, the small hand will be just after the 7, showing it is just after 7 o'clock. The big hand will be at 20 minutes, which is at 4.

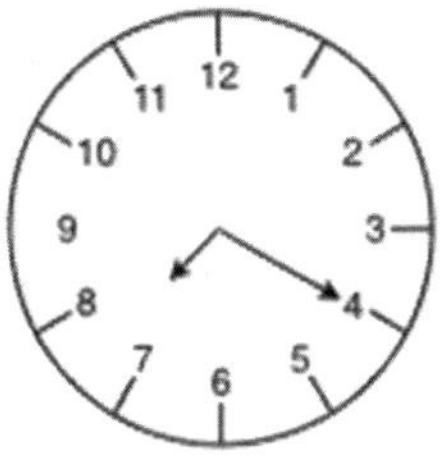

On a digital clock, the time will be written in the order: hours, then minutes. Since the hour is only one digit, there may be a 0 before the 7.

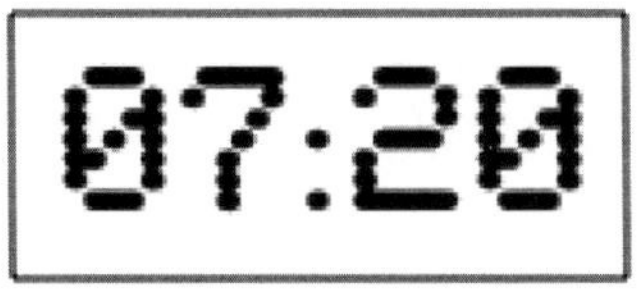

Find the time on the clock below.

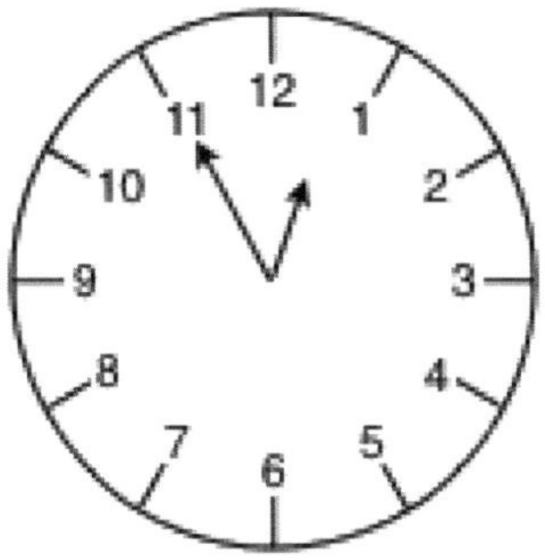

The big hand is on the 11, so the minutes is 55. The small hand is just before the 1, which means that it is close to 1 o'clock, but it is not yet 1 o'clock. It is still 12 o'clock. The time is 12:55.

Time Intervals

THINGS TO REMEMBER:

There are 60 seconds in 1 minute.

There are 60 minutes in 1 hour.

There are 24 hours in 1 day.

There are 365 days in 1 year.

ADDING AND SUBTRACTING TIME

When adding or subtracting time in minutes, an hour can be visualized as 4 sections of 15 minutes.

$\frac{1}{4}$ of an hour: 15 minutes

$\frac{2}{4}$ of an hour: 30 minutes

$\frac{3}{4}$ of an hour: 45 minutes

$\frac{4}{4}$ of an hour: 60 minutes

STRATEGIES FOR ADDING AND SUBTRACTING TIME

For example, if a student practices the piano for 15 minutes a day, for 4 days, how long have they practiced in all?

Use a model. Each 15-minute session represents $\frac{1}{4}$ of an hour. $15 + 15 + 15 + 15 = 60$. So, the student has practiced for 60 minutes, which is 1 hour.

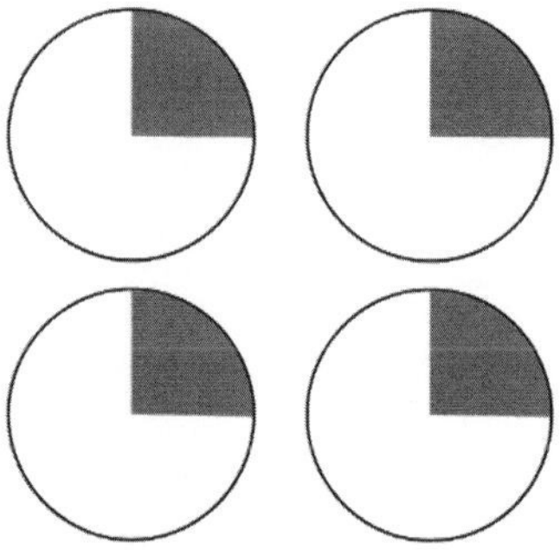

Using sections of the clock to add or subtract is a great way to solve problems mentally. Remember, one quarter of an hour is 15 minutes, and one half of an hour is 30 minutes. Skip counting by 15 or 30 is an efficient way to solve problems involving elapsed time. This also works for subtraction problems.

Consider the following scenario: A student studied 45 minutes for a math test, and 15 minutes for a science test. How much longer did the student study for the math test than the science test?

Math: 45 minutes

Science: 15 minutes

Subtract the time between studying for math and studying for science: $45 - 15 = 30$. This represents half an hour. The student spent 30 minutes more studying for the math test than the science test.

Temperature

Find the temperature, in degrees Fahrenheit, on the thermometer below:

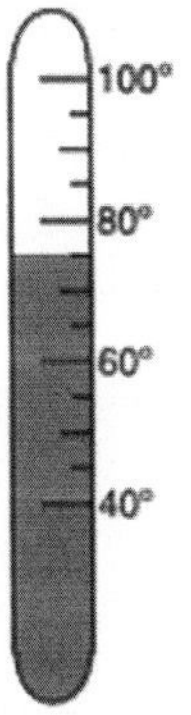

The temperature on the thermometer is halfway between 70º and 80º. The temperature halfway between 70º and 80º is 75º F.

Probability, Statistics, Patterns, Functions, and Algebra

Probability

Probability is the likelihood of a certain outcome occurring for a given event. An **event** is any situation that produces a result. It could be something as simple as flipping a coin or as complex as launching a rocket. Simple probability can be found using the formula below.

$$P = \frac{\text{Number of desired outcomes}}{\text{Number of total possible outcomes}}$$

EXAMPLE

Darla is going to make 24 cookies total. 5 of them will be chocolate chip, 15 of them will be sugar, and 4 of them will be snickerdoodles. If she randomly selected a cookie off the trays, what would be the likelihood that she selects a sugar cookie?

Since 15 out of the 24 total cookies are sugar cookies, the probability is $\frac{15}{24}$ which simplifies to $\frac{5}{8}$. The likelihood that she selects a sugar cookie is very high since majority of her cookies on the trays are sugar cookies.

Summarizing Data

Data refers to a collection of information. For example, when a survey is conducted, data points are collected. It is helpful to organize these data points before displaying the information in a table or graph. Typical data displays include bar graphs, frequency tables, pictographs, and dot plots.

BAR GRAPHS

Bar graphs use horizontal or vertical bars to represent the total amount for each category. For example, this bar graph shows the number of votes for a variety of snacks. The length of the bar represents the number of votes. Notice how the red popcorn bar represents 5 votes, the green crackers bar represents 2 votes, and so on. The interval should always be scaled and consistent when displaying data graphically. For example, the number of votes on this bar graph increases by 2 consistently.

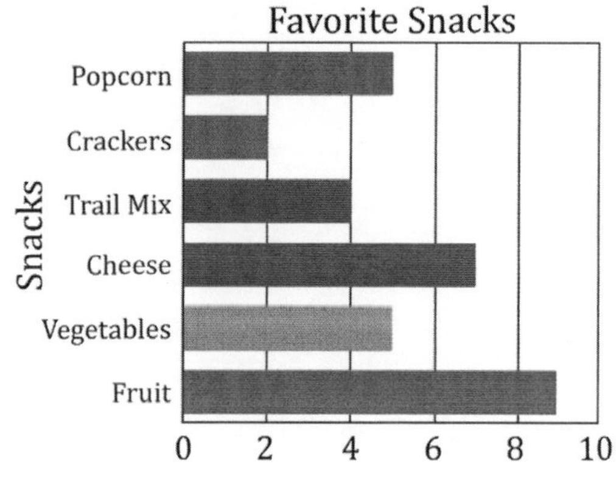

Frequency Tables

Frequency tables are used to show how often, or frequently, an event or value occurs. Frequency tables often use tally marks to show the total frequency for each category.

Favorite Club

Club	Tally	Frequency
Hockey	卌 |||	8
Badminton	卌 ||	7
Football	卌	5

Pictographs

Data can be summarized by studying graphical displays such as these. Consider the pictograph below. The **pictograph** uses pictures to represent the value for each category. In this pictograph, each picture represents 1 student. The pictograph shows the variety of ways students in a 3rd grade class commute to school each day. The graph can be used to summarize the data that was collected. For example, since each picture represents 1 student, we know there are 20 students in the class. We also know the most common mode of transportation is taking the bus, and the least common mode of transportation is walking on foot.

Ways to Get to School

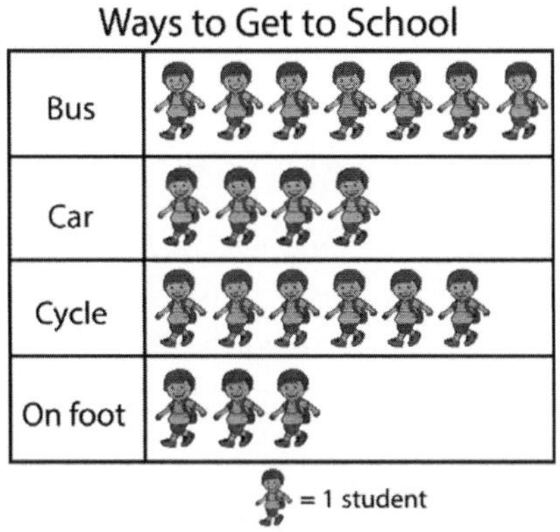

Review Video: Pictographs
Visit mometrix.com/academy and enter code: 147860

Example

Based on the information in the bar graph, how many more students prefer math or science than social studies or ELA?

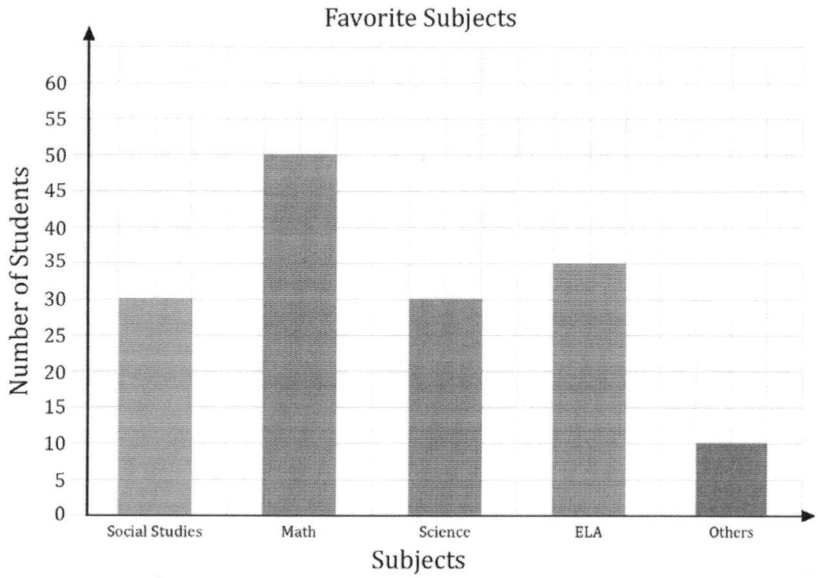

Students that prefer math or science: $50 + 30 = 80$.
Students that prefer social studies or ELA: $30 + 35 = 65$.
The difference: $80 - 65 = 15$.
There are 15 more students who prefer math or science than social studies or ELA.

Example

Which day of the week has a value of 30 cars?

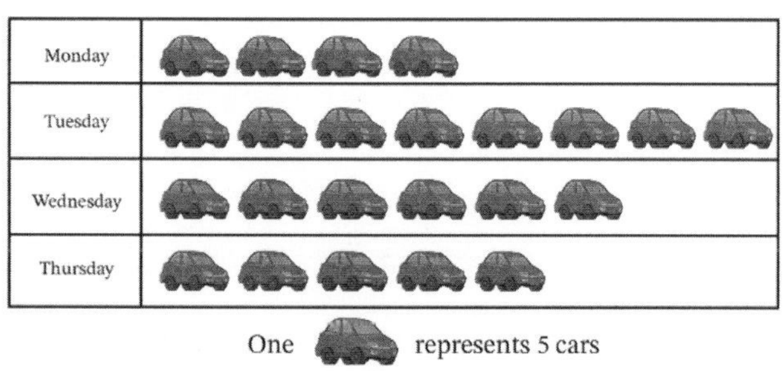

Each car in the pictograph represents 5 cars.
Monday shows 4 cars: $4 \times 5 = 20$ total cars.
Tuesday shows 8 cars: $8 \times 5 = 40$ total cars.
Wednesday shows 6 cars: $6 \times 5 = 30$ total cars.
Thursday shows 5 cars: $5 \times 5 = 25$ total cars.
Wednesday has a value of 30 total cars.

Patterns

A pattern is a particular arrangement of shapes, colors, pictures, numbers, objects, and other various things that repeats. Each pattern has a specific rule to follow. Patterns can also be referred to as sequences.

Example 1

Find the next number in each pattern:

1) 2, 4, 6, 8, ...

2) 5, 10, 15, 20, ...

3) 60, 50, 40, 30, ...

> 1) The numbers are increasing by 2. Add 2 to the last number to find the next number: 2, 4, 6, 8, 8 + 2 = 10.
>
> 2) The numbers are increasing by 5. Add 5 to the last number to find the next number: 5, 10, 15, 20, 20 + 5 = 25.
>
> 3) The numbers are decreasing by 10. Subtract 10 from the last number to find the next number: 60, 50, 40, 30, 30 – 10 = 20.

Example 2

Draw the next figure in this pattern:

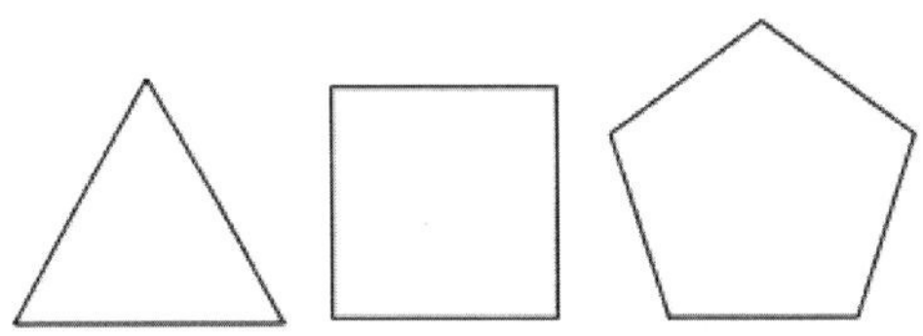

Look at the sides of each figure.

Figure 1: 3 sides

Figure 2: 4 sides

Figure 2: 5 sides

The number of sides on each figure is increasing by 1. The next figure will have 6 sides.

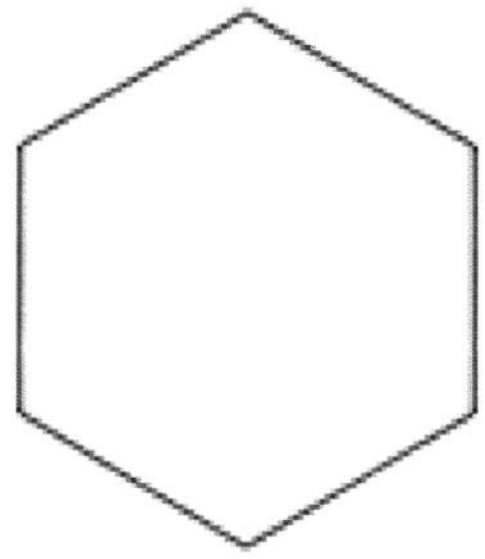

Example 3

Find the next figure in the pattern.

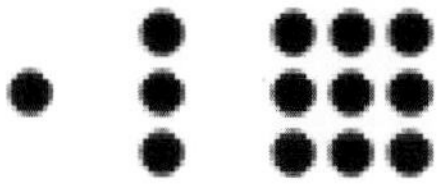

The first figure has 1 dot. The second has 3 dots. The third has 9 dots. Each figure is 3 times the size of the figure before it. The next figure will be three of the last figure:

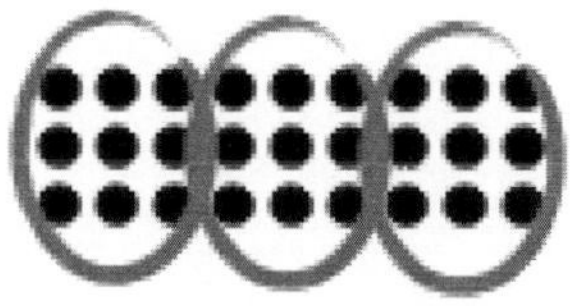

A figure that includes three of the last figure has $3 \cdot 9 = 27$ dots.

Finding Unknown Numbers

Inverse Operations

Multiplication and division go hand in hand. Division and multiplication are considered **inverse operations**, which means they are the reverse, or opposite, of each other. This can be helpful when there is an **unknown number** in an equation. For example, $4 \times ? = 24$ is a multiplication equation that can be solved using division. $24 \div 4 = 6$, therefore $4 \times 6 = 24$. The missing number is 6.

Let's try another example: $72 \div ? = 9$. This is a division equation that can be solved using multiplication. What number times 9 equals 72? The missing number is 8.

Fact Families

Don't forget to use **fact families** when solving equations with **missing numbers**.

For example, the equations $5 \times 8 = 40, 8 \times 5 = 40, 40 \div 5 = 8, 40 \div 8 = 5$ represent a fact family. The same three numbers are used in all the equations. This can be helpful when one of the numbers is missing. For example, $8 \times ? = 40$. Remember the fact families. 8 groups of 5 equals 40, so the missing number is 5.

$$\text{Factor} \times \text{Factor} = \text{Product}$$

$$8 \times 5 = 40$$

8 and 5 are considered factors, and 40 is the product.

Equivalent Mathematical Relationships

Equations can be used to represent equivalent mathematical relationships. Equations consist of two equivalent expressions joined together by an equal sign.

Example

Write and solve a subtraction equation that matches the model below.

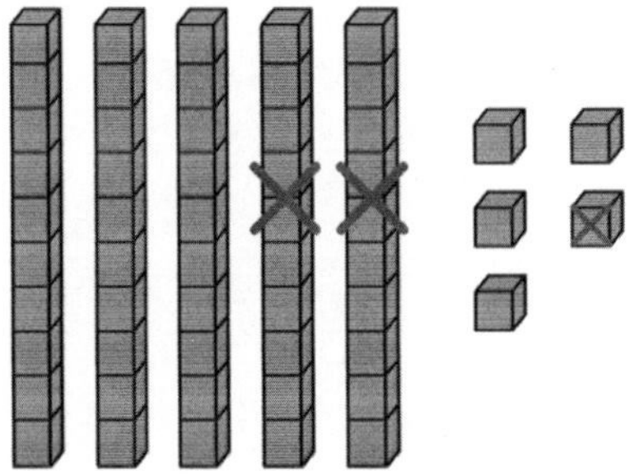

The model shows 5 groups of ten plus 5 groups of one, which represents 55. Then 2 groups of ten and 1 group of one are crossed out. This represents subtraction of 21. This leaves 3 groups of ten and 4 groups of one. So, $55 - 21$ leaves 34 remaining. The answer is $55 - 21 = 34$.

Example

Write two multiplication equations the array below can represent.

The array shows rows and columns of stars. This can be described as 5 rows of 6, or 6 rows of 5. Both scenarios have a product of 30.

The answers are $5 \times 6 = 30$ and $6 \times 5 = 30$.

EXAMPLE

Use the numbers 3, 6, and 18 to write four equations using multiplication and division.

The numbers 3, 6, and 18 form a fact family.

3 groups of 6 equals 18: $3 \times 6 = 18$.

6 groups of 3 equals 18: $6 \times 3 = 18$.

18 divided into 3 equal groups results in 6 per group: $18 \div 3 = 6$.

18 divided into 6 equal groups results in 3 per group: $18 \div 6 = 3$.

The answers are $3 \times 6 = 18$, $6 \times 3 = 18$, $18 \div 3 = 6$, $18 \div 6 = 3$.

Practice Test #1

1. Arlan compares his annual electricity expenses over a five-year time span. His annual expenses are shown in the table below.

Year	Expense
2007	$1,224
2008	$1,319
2009	$1,046
2010	$1,529
2011	$1,342

Which of the following shows the years listed, in order, from lowest electricity expense to highest electricity expense?

a. 2009, 2008, 2007, 2010, 2011
b. 2007, 2009, 2011, 2010, 2008
c. 2009, 2007, 2011, 2008, 2010
d. 2009, 2007, 2008, 2011, 2010

2. Steve had to pay a library fee. The amount he paid is shown below.

How much did he pay?

a. $2.84
b. $2.59
c. $2.74
d. $2.69

3. Wyatt's coach passed out balls to the entire team. One-third of the balls Wyatt received were basketballs. Which of the following could represent the balls he received?

a 

b. 

c. 

d. 

4. An author had a book published last year. The number of books sold each month, for the first six months are: 321, 452, 608, 381, 294, and 458.

Which of the following lists the number of books sold each month, in order from greatest to least?

a. 608, 452, 458, 321, 384, 294
b. 452, 458, 608, 381, 321, 294
c. 458, 608, 452, 381, 321, 294
d. 608, 458, 452, 381, 321, 294

5. Students were given triangles and hexagons to be used during a class activity. One-third of the shapes each student received were triangles. Which of the following could represent the number of triangles and number of hexagons each student received?

a. 

b. 

c. 

d. 

6. What number sentence is illustrated by the diagram below?

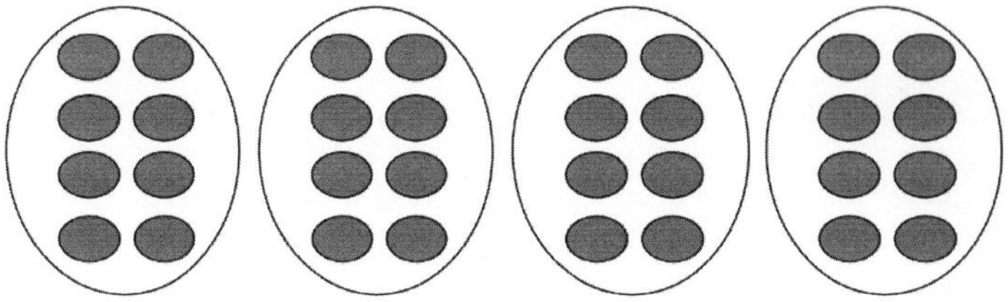

a. $32 \times 4 = 128$
b. $32 - 4 = 28$
c. $32 \div 4 = 8$
d. $32 + 4 = 36$

7. Solve and place answer on the provided griddable sheet.

$$\begin{array}{r} 96 \\ \underline{\times 5} \end{array}$$

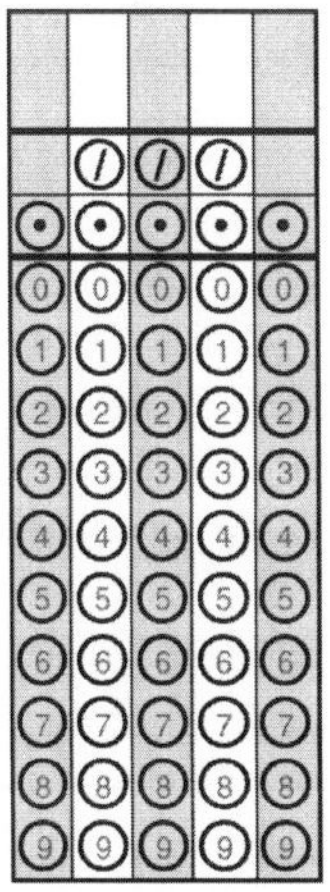

8. Edward ascends to the top of a mountain over the course of two days. On Friday, he ascends 482 feet. He ascends another 362 feet on Saturday. How many feet did he ascend in all?

a. 848 feet
b. 836 feet
c. 840 feet
d. 844 feet

9. Kristen must buy two items that are priced at $62, and $89. What is the best estimate for the total cost of both items?

a. $120
b. $150
c. $200
d. $180

10. Which of the following represents $\frac{2}{3}$?

a. 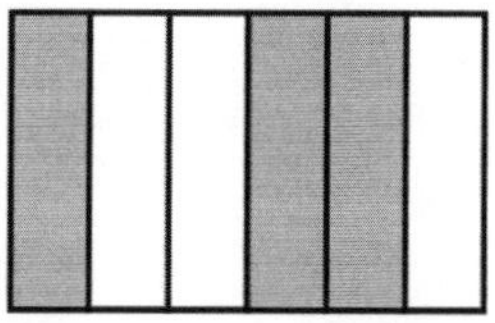

b. 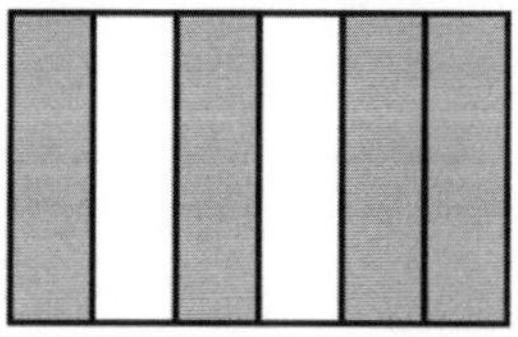

c. 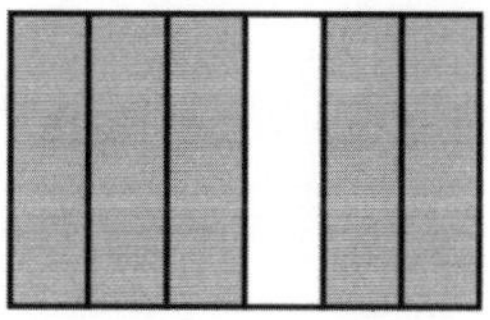

d. 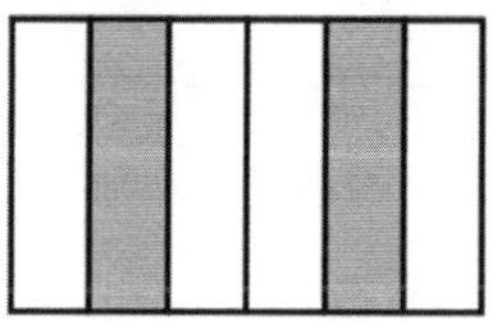

11. On Monday, David finished gluing 96 tiles in 8 hours. On Tuesday, he finished gluing 72 tiles in 8 hours. Which of the following is a possible first step in determining how many more tiles he glued per hour on Monday?

a. Add the number of tiles glued each day
b. Subtract the number of hours it took to glue the tiles from the number of tiles glued
c. Multiply the number of hours spent gluing tiles on Monday by the number of hours spent gluing tiles on Tuesday
d. Divide the number of tiles glued each day by the number of hours it took to glue them

12. Solve and place answer on the provided griddable sheet.

At the birthday party, each guest received 12 game tokens and each game takes 2 tokens. If there were 9 guests at the party, estimate how many games were played (to the nearest tens)?

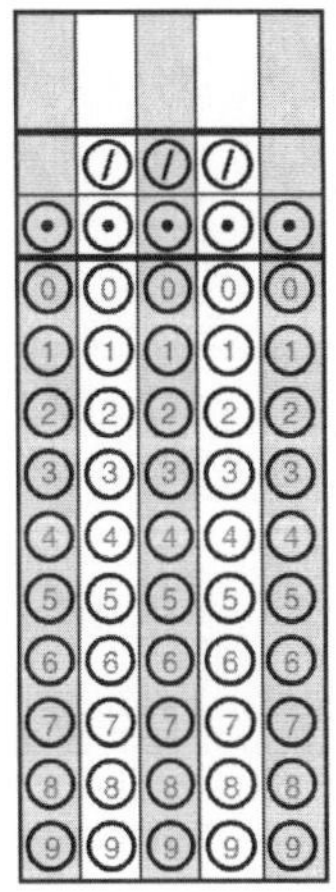

13. What is the eighth number in the pattern: 321, 329, 337, 345, ___, ___, ___, ___?

a. 369
b. 385
c. 377
d. 353

14. A little boy has three nickels, four dimes, two pennies, and two quarters. How much money does he have?

a. $1.00
b. $0.90
c. $1.07
d. $100

15. What is the value of the underlined digit in 23,085?

a. 8000
b. 800
c. 80
d. 8

16. Solve and place answer on the provided griddable sheet.

One number can correctly fill in all the blank boxes below. What is that number?

$$\_\_\_ \times 3 = 12$$
$$16 - \_\_\_ = 12$$
$$8 + \_\_\_ = 12$$

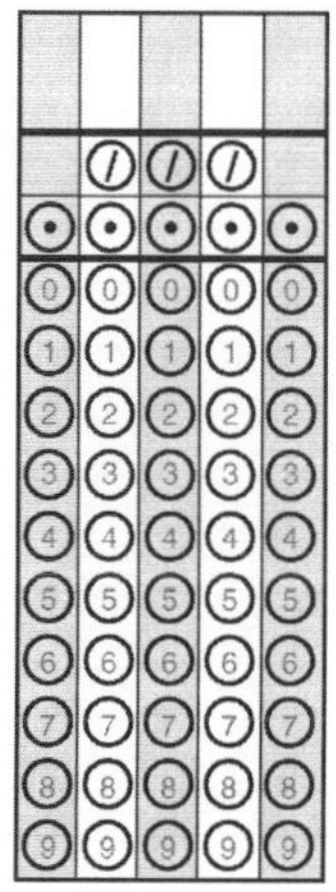

17. A class collects spiders. Spiders have 8 legs each. Which table shows the number of legs found on the spiders brought to class?

a.

Number of Spiders	Number of Legs
2	10
3	11
6	14
8	16
11	19

b.

Number of Spiders	Number of Legs
4	28
5	35
9	63
12	84
14	98

c.

Number of Spiders	Number of Legs
3	12
4	13
8	17
10	19
12	21

d.

Number of Spiders	Number of Legs
3	24
5	40
6	48
9	72
12	96

18. The total number of candy pieces found in different numbers of candy jars is shown in the table below.

Number of Candy Jars	Number of Candy Pieces
2	28
3	42
4	56
5	70

How many candy pieces are there in 8 candy jars?

a. 84
b. 126
c. 112
d. 98

19. Penny drinks 8 glasses of water each day. The number of glasses of water she drinks over a 7-day time span can be determined, using the number sentence: $8 \times 7 = ?$

Which number sentence would not show the number of glasses of water she drinks?

a. $? \div 8 = 7$
b. $7 \times 8 = ?$
c. $? \div 7 = 8$
d. $7 - 8 = ?$

20. Belinda draws a rectangle with a length of 6 cm. She draws a second rectangle with a length of 11 cm. Belinda continues drawing more rectangles, where for each rectangle drawn, she uses a length that is 5 more centimeters than the length of the previous rectangle. If this pattern continues, what will be the length of the 11th rectangle?

a. 46 cm
b. 54 cm
c. 56 cm
d. 61 cm

21. What is the next number in the sequence shown?

132, 123, 115, 108, 102

a. 82
b. 87
c. 92
d. 97

22. Which of the following figures is NOT congruent to the others shown?

a. 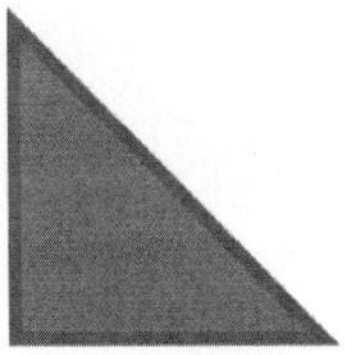

b. 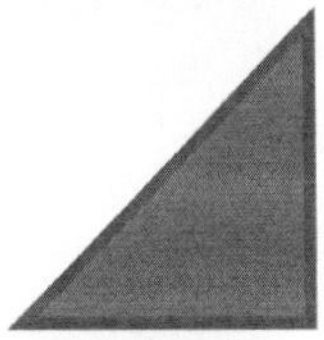

c. 

d. 

23. Which number is greater than the number shown by Point M on the number line?

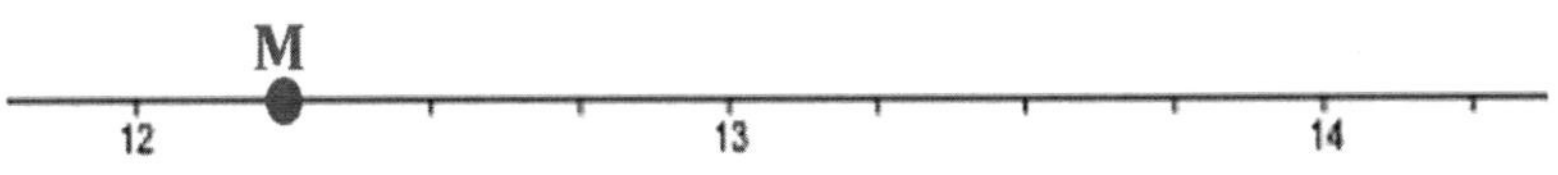

a. $12\frac{1}{4}$
b. $12\frac{1}{2}$
c. 12
d. $12\frac{3}{4}$

24. What is 788 rounded to the nearest ten?

a. 789
b. 800
c. 790
d. 780

25. Which of the following is true?

a. $\frac{3}{5} = \frac{4}{5}$
b. $\frac{3}{5} < \frac{4}{5}$
c. $\frac{2}{5} > \frac{4}{5}$
d. $\frac{2}{5} = \frac{4}{5}$

26. Which shape has the most sides?

a. Square
b. Triangle
c. Pentagon
d. Hexagon

27. In which place value is the digit 3 in 4,382?

a. Ones
b. Tens
c. Hundreds
d. Thousands

28. Find the sum of $\frac{12}{25} + \frac{7}{25}$.

a. $\frac{19}{50}$
b. $\frac{1}{5}$
c. $\frac{19}{25}$
d. $\frac{4}{5}$

29. What number does Point P represent?

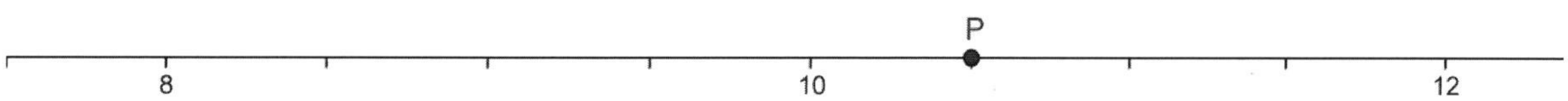

a. $10\frac{1}{4}$
b. $10\frac{1}{2}$
c. $10\frac{1}{3}$
d. $10\frac{3}{4}$

30. What is 125,420 rounded to the nearest hundred?

a. 400
b. 500
c. 125,000
d. 125,400

31. What is the perimeter of the triangle shown below?

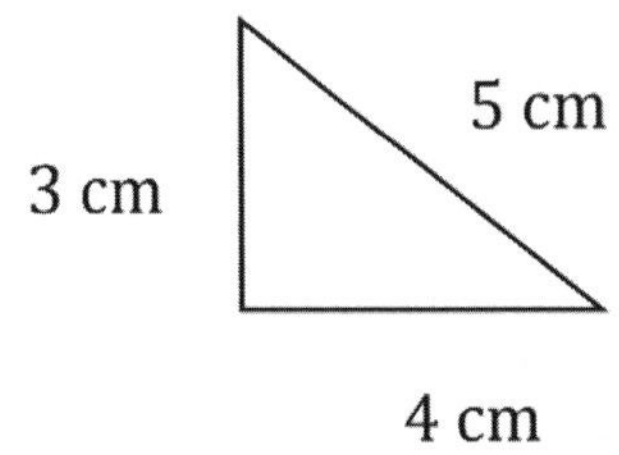

a. 6 cm
b. 12 cm
c. 9 cm
d. 8 cm

32. Which operation symbol properly compares $\frac{2}{3}$ and $\frac{5}{6}$?

a. >
b. =
c. <
d. Impossible to determine

33. Order these numbers from greatest to least.

4,586	7,324	9,126	3,854	4,789	9,087

a. 9,126; 7,324; 9,087; 4,568; 4,789; 3,854
b. 9,126; 9,087; 7,324; 4,789; 4,568; 3,854
c. 3,854; 4,568; 4,789; 7,324; 9,087; 9,126
d. 9,126; 9,087; 7,324; 4,568; 4,789; 3,854

34. A pencil is shown below. Use the ruler from your mathematics chart to measure the length of the pencil. About how long is the pencil, from base to tip, to the nearest inch?

a. 2 in
b. 3 in
c. 4 in
d. 5 in

35. How many square units are found in the trapezoid below?

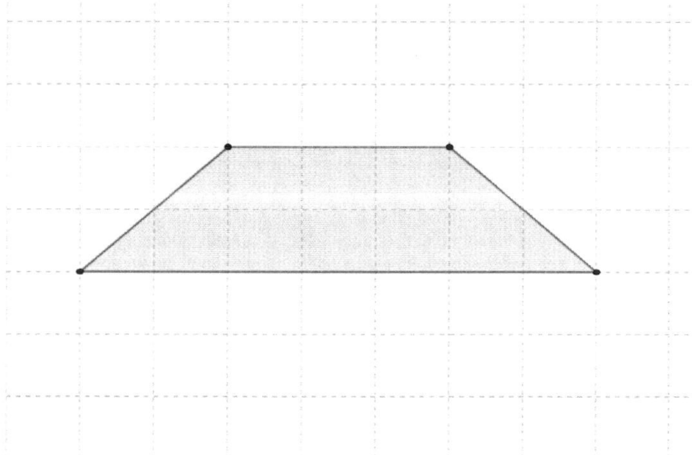

a. 8 square units
b. 12 square units
c. 10 square units
d. 9 square units

36.

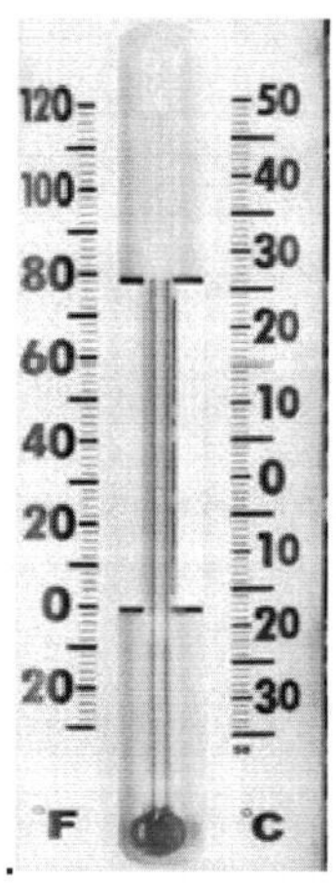

What is the approximate temperature, in degrees Fahrenheit?

a. 0° F
b. 80° F
c. 30° F
d. 25° F

37. Karen goes to the library at the time shown on the clock below. At what time does she go to the library?

a. 8:10
b. 1:30
c. 6:10
d. 2:30

38. Which clock below would be read as a five o'clock?

a. **5:00**

b. **5:30**

c. **5:05**

d. **5:55**

39. Aubrey rolls a die. Which of the following statements is true?

a. She is more likely to roll a 3 than a 6
b. She is equally likely to roll a 3 or a 6
c. She is less likely to roll a 3 than a 6
d. She is certain to roll a 3 or a 6

40. The bar graph below shows the number of zoos found in four states.

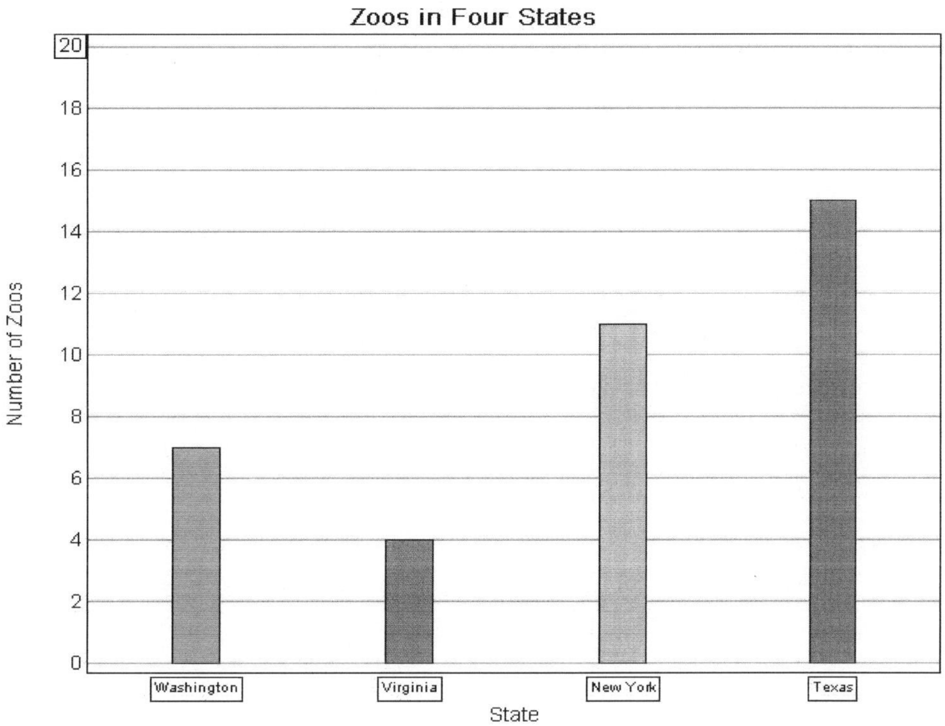

Which state has the most zoos?

a. Washington
b. Virginia
c. New York
d. Texas

41. A fish bowl contains 2 striped, 4 orange, and 2 blue fish. Eli randomly scoops a fish from the bowl. Which of the following statements is true?

a. He is less likely to scoop an orange than a striped fish
b. He is more likely to scoop a striped than a blue fish
c. He is more likely to scoop a blue than an orange fish
d. He is equally likely to scoop a striped and a blue fish

42. The number of lawns, finished by four different landscaping companies, in one week, is shown in the table below.

Landscaping Company	Number of Lawns
Landscaping Company A	12
Landscaping Company B	18
Landscaping Company C	6
Landscaping Company D	16

Which pictograph shows the number of lawns finished by each company?

a.
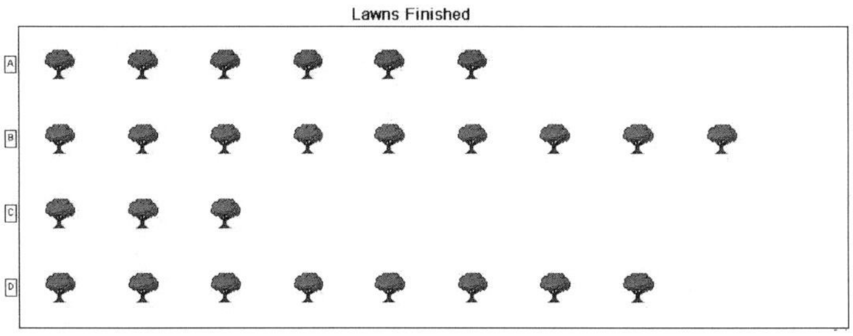

Each tree represents 2 lawns.

b.
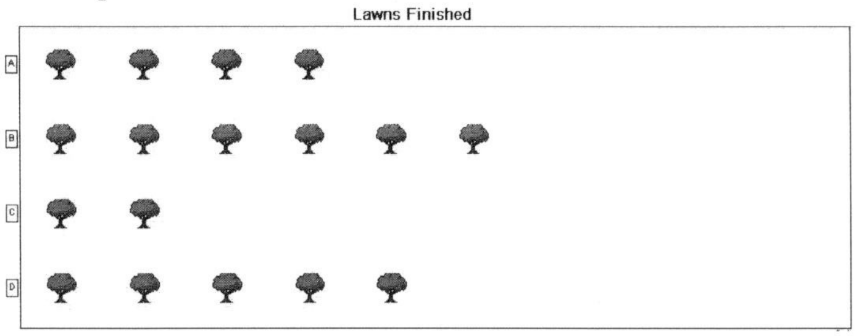

Each tree represents 4 lawns.

c.
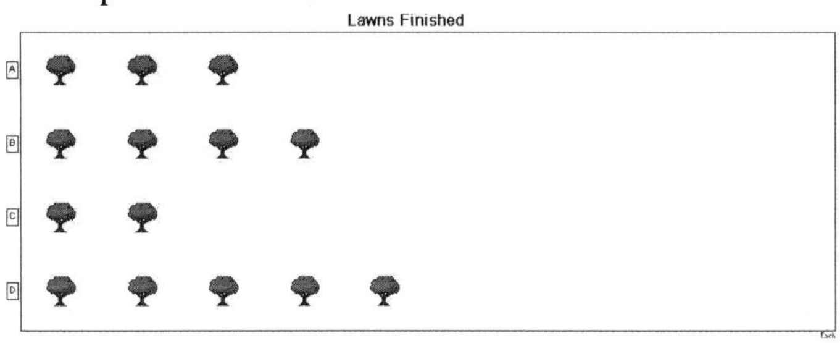

Each tree represents 2 lawns.

d.
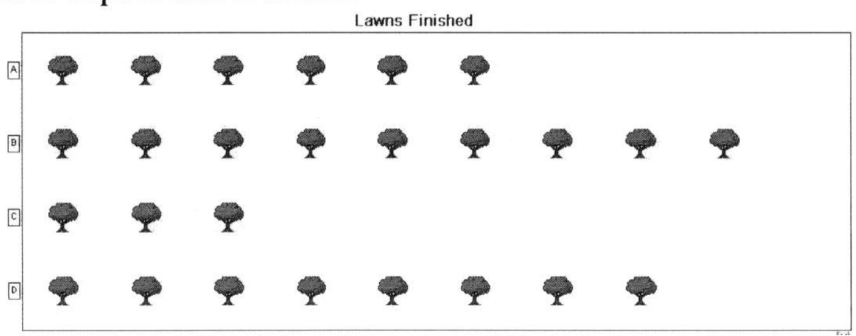

Each tree represents 3 lawns.

43. Which of the following events is certain?

a. Rolling a 7 on a die
b. Rolling an even number on a die
c. Rolling a number less than 8 on a die
d. Rolling an odd number on a die

44. The bar graph below shows the favorite sport of the students in the class.

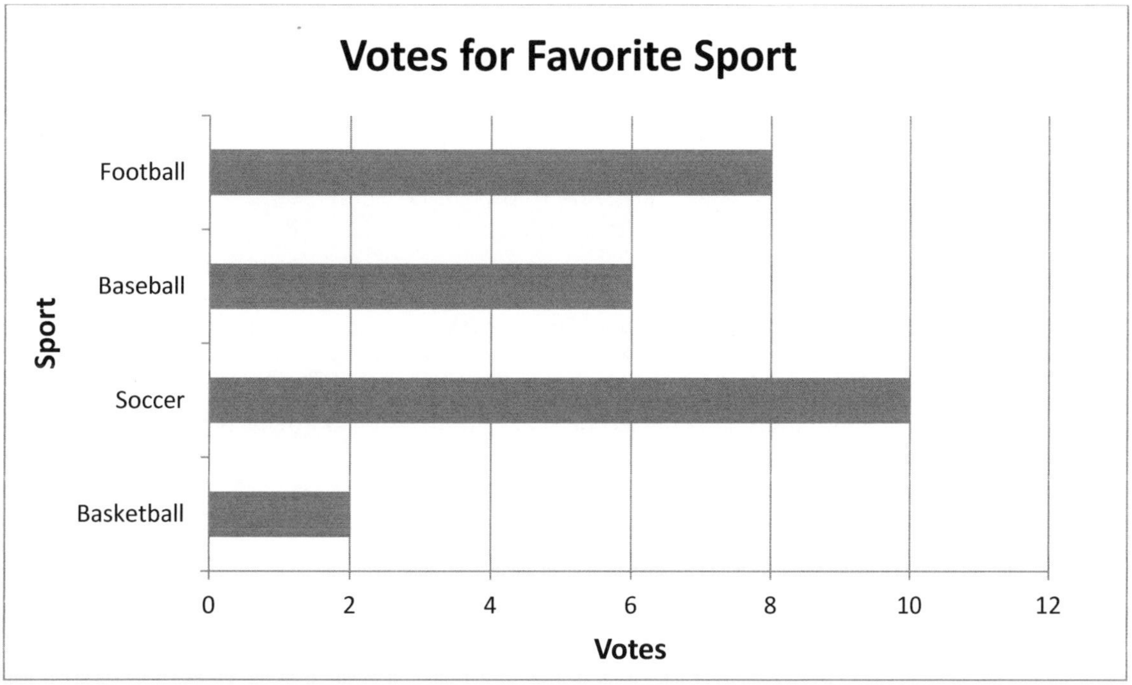

How many fewer votes did basketball get compared to baseball?

a. 2
b. 3
c. 4
d. 5

Answer Key and Explanations for Test #1

1. D: The annual expenses for Years 2009, 2007, 2008, 2011, and 2010 are \$1,046, \$1,224, \$1,319, \$1,342, and \$1,529, respectively. These amounts are listed in order from lowest to highest. Since all of the numbers have a 1 in the thousands place, the numerals in the hundreds place must be compared. For the amounts of \$1,319 and \$1,342, the numerals in the tens place must be compared. No other choice shows the years listed in increasing order of expense.

2. C: There are 2 dollar bills, which represent 2 dollars. There are also 2 quarters, 2 nickels, 1 dime, and 4 pennies. Two quarters are worth \$0.50 since each is worth \$0.25 $(2 \times 0.25 = 0.50)$, 2 nickels are worth \$0.10 since each is worth \$0.05 $(2 \times 0.05 = 0.10)$, 1 dime is worth \$0.10, and 4 pennies are worth \$0.04 since each is worth \$0.01. The sum of the coins can be found by writing: $\$0.50 + \$0.10 + \$0.10 + \$0.04$, which equals \$0.74. The sum of the two dollar bills and the coins can be written as: $\$2.00 + \$0.74$. Thus, he paid \$2.74.

3. B: If $\frac{1}{3}$ of the balls Wyatt received were basketballs that means the remaining balls he received must have been soccer balls. Since the fraction used is $\frac{1}{3}$, look at it as if he received 3 balls. One ball was a basketball and the remaining balls had to be soccer balls. This is now a simple subtraction problem. $3 - 1 = 2$ soccer balls.

4. D: The numbers: 608, 458, 452, 381, 321, and 294 are written in order from greatest to least. An examination of the numerals in the hundreds place reveals that 608 is the greatest and 294 is the least. The numbers, 458 and 452, are greater than the numbers, 381 and 321, based on the value of the digits in the hundreds place. An examination of the numerals in the ones place, since both the digits in the hundreds and the tens places are the same, allows you to compare 458 and 452, which shows 458 to be the larger number. An examination of the numerals in the tens place allows you to compare 381 and 321, since the digits in the hundreds place are the same, which shows 381 to be the larger number. None of the other choices show the numbers in descending order.

5. C: Choice C shows 2 triangles and 4 hexagons, which is a total of 6 shapes. The fraction 2 out of 6, or $\frac{2}{6}$, also represents 1 out of 3, or $\frac{1}{3}$. None of the other choices represent this fraction: Choice A shows $\frac{3}{6}$, which equals $\frac{1}{2}$. Choice B shows $\frac{1}{4}$. Choice D shows $\frac{1}{6}$.

6. C: The diagram shows 32 counters divided into 4 groups, with 8 counters in each group. Therefore, the total number of counters, 32, is divided by 4, giving a quotient of 8, which is written as: $32 \div 4 = 8$.

7. 480: This is a simple multiplication problem which involves carrying the excess (greater than a single digit) to the next column. Start with the 5 and multiply each digit above: $5 \times 6 = 30$. Write down the 0 and place the 3 on top of the 9. Multiply 5×9 to get 45. Add in the 3 to get 48. Write down the number 48 and the product is 480.

8. D: The total number of feet he ascended can be determined by adding 482 feet and 362 feet. The sum of 482 and 362 is 844. Thus, he ascended 844 feet in all.

9. B: The item priced at \$62 can be rounded to \$60. The item priced at \$89 can be rounded to \$90. The sum of 60 and 90 is 150. Thus, the best estimate is \$150.

10. B: Choice B shows 4 shaded sections out of 6 total sections. The fraction, $\frac{4}{6}$, is the same as the fraction, $\frac{2}{3}$. Two shaded sections represent one-third of the total. Thus, four shaded sections represent two-thirds of the total. Each of the pictures has 6 total sections, so the other choices can be written as fractions with a 6 in the denominator. Choice A shows $\frac{3}{6}$, which equals $\frac{1}{2}$. Choice C shows . Choice D shows $\frac{2}{6}$, which equals $\frac{1}{3}$. So, only Choice B shows the correct picture.

11. D: A possible first step would be to divide the number of tiles glued each day by the number of hours it took to glue the tiles. The two quotients would then represent approximately how many tiles glued per hour, and could then be compared.

12. 50: Begin with the tokens given to each guest, which was 12. But the problem states that every game takes two tokens, so each guest can play only 6 games, not 12. Multiply this by the number of guests, 9, to equal 54 games. Round this number to the nearest tens to get 50.

13. C: Each number is 8 more than the previous number, which means that the next number can be found by adding 8 to the previous number. Thus, the fifth number is 353, the sixth number is 361, the seventh number is 369, and the eighth number is 377.

14. C: Convert the coins to cents and apply addition. Each nickel is worth 5 cents, so 3 nickels equals 15 cents. Each dime is worth 10 cents, so 4 dimes equals 40 cents. Each penny is worth 1 cent, so 2 pennies equals 2 cents. A quarter is worth 25 cents, so 2 quarters equals 50 cents. $15 + 40 + 2 + 50 = 107$ cents. Since there are 100 cents in a dollar, this becomes 1 dollar with 7 cents remaining or $1.07

15. C: The place value is assigned by the position of the digit. The places from left to right are ones, tens, hundreds, thousands, ten-thousands, etc. Since the digit 8 is in the second place from the left, it is in the tens place. The value of the digit is 8×10, which is 80.

16. 12: The correct answer is 4. $4 \times 3 = 12$; $16 - 4 = 12$, and $8 + 4 = 12$.

17. D: Each spider has 8 legs. In order to find the number of legs present with 3 spiders, you multiply 3 by 8, which is 24. Thus, 3 spiders have 24 legs in all. Choice D is the only table that shows each number of spiders, multiplied by 8, to yield the correct product representing the total number of legs.

18. C: Each candy jar has 14 pieces of candy. This can be determined by dividing the number of pieces of candy by the number of candy jars: $28 \div 2 = 14, 42 \div 3 - 14, 56 \div 4 = 14, 70 \div 5 = 14$. Since the data in the table shows that there are 14 pieces of candy in each jar, multiplying $8 \times 14 = 112$ finds the total number of pieces of candy that are in 8 candy jars.

19. D: If she drinks 8 glasses of water each day, the number of glasses of water she drinks in 12 days can be determined by multiplying 8 by 7. This product is 56; thus she drinks 56 glasses of water in a 7-day time span. The relationship between the number of glasses of water she drinks per day and the total number of glasses of water she drinks in 7 days can be represented by an appropriate multiplication or division number sentence within the following fact family: $8 \times 7 = 56, 56 \div 8 = 7, 7 \times 8 = 56, 56 \div 7 = 8$. Subtracting 8 from 7 will not reveal the number of glasses she drinks in a 7-day time span. The number sentence: $7 - 8 = ?$, is not in this fact family.

20. C: If each rectangle has a length that is 5 cm more than that of the previous rectangle, the lengths of the 3rd rectangle through the 11th rectangle can be found by adding 5 cm to the length of the second triangle and continuing for each next rectangle. So, the lengths of the rectangles will be as follows: 16 cm, 21 cm, 26 cm, 31 cm, 36 cm, 41 cm, 46 cm, 51 cm, and 56 cm.

21. D: The pattern is subtracting one less number each time:

$$132 - 9 = 123$$
$$123 - 8 = 115$$
$$115 - 7 = 108$$
$$108 - 6 = 102$$

The next number to be subtracted is 5, so $102 - 5 = 97$.

22. D: Choices A, B, and C are all the same right triangle just flipped around. Answer choice D is a different triangle and therefore not congruent.

23. D: $12\frac{1}{4}$ is represented by the first tick mark to the right of the 12. $12\frac{1}{2}$ is the second tick mark and $12\frac{3}{4}$ is the third tick mark, or the tick mark before the 13. Only $12\frac{3}{4}$ appears to the right of point M on the number line, making it greater.

24. C: The place value of a digit depends on its location in relation to the decimal point. The places to the left of the decimal are ones, tens, then hundreds, respectively. To round a number to a certain place value, look at the digit to the immediate right. If the digit to the right is 5 or more, round up by adding one. If the digit is less than five, round down by keeping the same digit. For this problem, the digit in the tens place is 8. The digit to its immediate right is 8. Since 8 is greater than 5, round up: 788 rounded to the nearest ten is 790.

25. B: $\frac{3}{5}$ is less than $\frac{4}{5}$. They are not equal. When comparing fractions with the same denominator, compare only the values in the numerator. Since they share a common denominator of 5, compare only the 3 and 4. Since 3 is less than 4, the same can be said for the fraction $\frac{3}{5}$ when compared to $\frac{4}{5}$.

26. D: A hexagon has 6 sides. A square has 4 sides. A triangle has 3 sides. A pentagon has 5 sides. Therefore, a hexagon has the most number of sides out of the options provided.

27. C: The place value is assigned by the position of the digit. The places from left to right are ones, tens, hundreds, thousands, ten-thousands, etc. The 3 is located in the hundreds place since it is three spaces from the left.

28. C: When adding fractions together that share a common denominator, add the numerators together and leave the denominator the same. Since both fractions share a denominator of 25, the numerators can be added together. 12 plus 7 is equal to 19. Therefore, $\frac{12}{25} + \frac{7}{25} = \mathbf{\frac{19}{25}}$.

29. B: Each increment represents one-half. This can be determined by counting that there are 3 marks, or 4 spaces, that lie between the difference of two wholes, as in between 10 and 12. Thus, one increment past 10, where Point P is located, represents $10\frac{1}{2}$.

30. D: The place value of a digit depends on its location in relation to the decimal point. The places to the left of the decimal are ones, tens, then hundreds, respectively. To round a number to a certain place value, look at the digit to the immediate right. If the digit to the right is 5 or more, round up by adding one followed by a zero. If the digit is less than five, round down by keeping the same digit followed by a zero. For this problem, the digit in the hundreds place is 4. The digit to its immediate right is 2. Since 2 is less than 5, round down. So, 125,420 rounded to the nearest hundred is 125,400.

31. B: The distance around the triangle, or the perimeter, is equal to the sum of 3 cm, 5 cm, and 4 cm. Thus, the perimeter is 12 cm.

32. C: One way to compare two fractions is by making two strip diagrams. The denominator represents how many equal spaces the strip will be divided into. The numerator represents how many spaces on the strip will be shaded in. It is important to make the strip diagrams the same size to accurately compare the fractions. When looking at the strip diagrams, it is evident that $\frac{2}{3}$ is less than $\frac{5}{6}$.

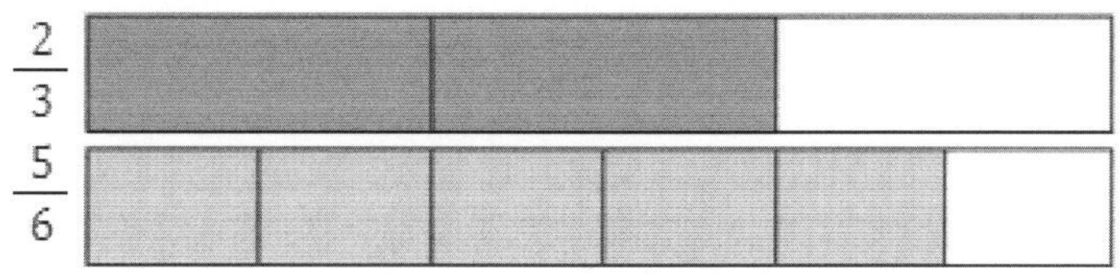

33. B: The correct order when order from greatest to least is 9,126; 9,087; 7,324; 4,789; 4,568; 3,854. 9,126 is the furthest to the right on the number line, so it is the largest. 3,854 is the furthest to the left on the number line, so it is the smallest.

34. B: The pencil is approximately 3.25 inches, which rounds to 3 inches.

35. C: The trapezoid has 8 square units, plus 4 one-half square units, which equals 2 more square units. The sum of 8 square units and 2 square units is 10 square units.

36. B: The thermometer shows the temperature to be very close to 80 degrees Fahrenheit. It also shows the temperature in Celsius: about 25 degrees.

37. C: The short hand, or hour hand, is between 6 o'clock and 7 o'clock, revealing that Karen went to the library after 6 o'clock. The long hand, or minute hand, is pointed at the 2. Since the 2 on the clock represents 10 minutes after the hour (since each number shown on the clock represents 5 minutes and $2 \times 5 = 10$), the clock shows that Karen went to the library at 6:10.

38. A: Five o'clock shown on a digital clock would be shown as a 5 followed by two zeroes, indicating no minutes.

39. B: Since a die has one of each number, from 1 to 6, she is equally likely to roll any of the six numbers. The possibility of rolling one of these numbers is no more or less than the possibility of rolling another of these numbers. Thus, she is equally likely to roll a 3 or a 6.

40. D: Texas has the most zoos because it has 15 zoos, while the other states each have 7 zoos, 4 zoos, and 11 zoos. Also, it can be seen from the graph that the bar representing Texas is much higher than the bars for the other states.

41. D: The more fish there is of a certain color the more likely it is that the color of fish is scooped. With more orange fish than striped fish in the bowl, he is more likely to scoop an orange fish than a striped one. There are equal amounts of striped and blue fish, so one is not more likely than the other. There are more orange fish than blue fish, so he is more likely to scoop an orange fish than a blue one.. Finally, the number of striped and blue fish is the same – so they are equally likely to be scooped compared to each other. Thus, Choice D is the only true statement.

42. A: Since each tree represents 2 lawns, the pictograph shows that the number of lawns finished by Company A is equal to 6×2, or 12 lawns, the number of lawns finished by Company B is equal to 9×2, or 18 lawns, the number of lawns finished by Company C is equal to 3×2, or 6 lawns, and the number of lawns finished by Company D is equal to 8×2, or 16 lawns. This is the only pictograph that represents the correct number of lawns.

43. C: A die has numbers 1-6, so rolling a number less than 8 is a certain occurrence. It will always happen. Choice A is incorrect because rolling a 7 would be impossible. Choices B and D are also incorrect because rolling either an even or odd number, since the die has both, would not be certain.

44. C: Baseball received 6 votes and basketball received only 2 votes. The difference is $6 - 2 = 4$ fewer votes.

Practice Test #2

1. What is the value of the underlined digit in 4$\underline{5}$,162?

a. 5000
b. 500
c. 50
d. 5

2. Which numbers are listed in order from least to greatest?

a. 539, 542, 528, 576, 588
b. 588, 576, 542, 539, 528
c. 528, 542, 539, 588, 576
d. 528, 539, 542, 576, 588

3. Which of the following is equations represents five subtracted from twenty-five equals twenty?

a. $25 - 20 = 5$
b. $5 - 25 = 5$
c. $25 - 5 = 20$
d. $20 + 5 = 25$

4. Camille's teacher passed out crayons and pencils to students in her class. One-fourth of the writing tools Camille received were crayons. Which of the following could represent the number of crayons she received?

a. 

b. 

c. 

d. 

5. Bercu sells 128 hot dogs this month. She sold 117 hot dogs last month. How many hot dogs has she sold in these past two months?

a. 235
b. 241
c. 242
d. 245

6. Chandler paid for a book at a book fare. He gave the clerk the amount of money shown below. How much money did he give the clerk?

a. $2.57
b. $2.72
c. $2.77
d. $2.82

7. What number sentence is shown by the diagram below?

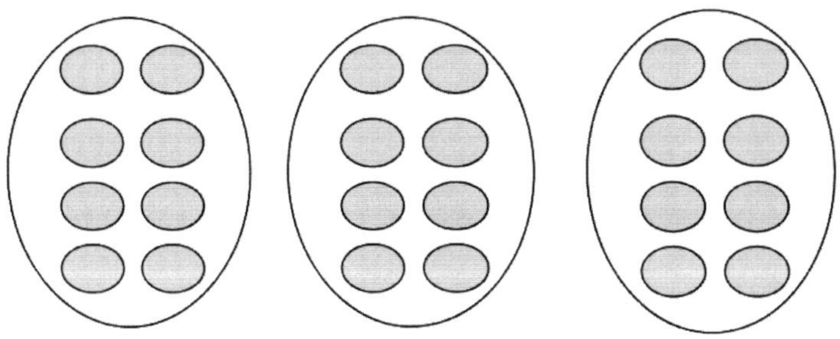

a. $24 \times 3 = 72$
b. $24 - 8 = 16$
c. $24 + 8 = 32$
d. $24 \div 3 = 8$

8. Solve and place answer on the provided griddable sheet.

$$\begin{array}{r} 78 \\ \underline{\times\ 4} \end{array}$$

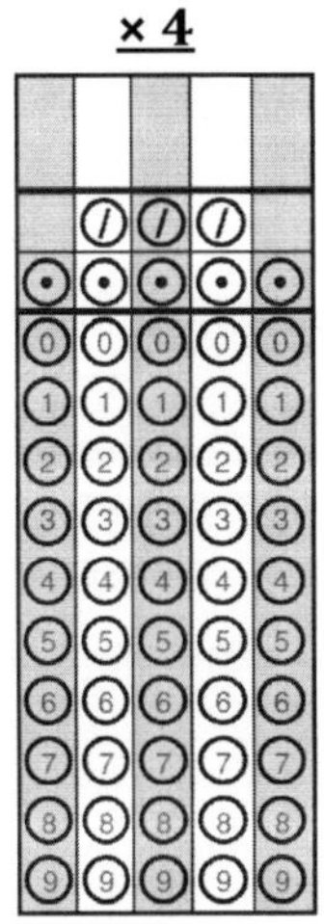

9. On Monday morning, a store manager gave 3 coupons to every customer who entered the store. If 89 people entered the store that morning, which is the best estimate for the total number of coupons the store manager gave to customers?

a. 300
b. 270
c. 480
d. 180

10. Farm A has 8 chickens and 3 horses. Farm B has 6 chickens and 5 horses. Which of the following is a possible first step in determining which farm contains more animal feet?

a. Multiply the number of chickens on each farm by 2, and multiply the number of horses on each farm by 4
b. Find the total number of animals found on both farms
c. Multiply the sum of the number of chickens and horses, found on both farms, by 6
d. Add 2 to the number of chickens found on each farm, and add 4 to the number of horses found on each farm

11. Which model shows a fraction that is more than 4 out of 7?

a. 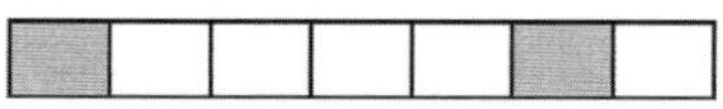
b. 
c. 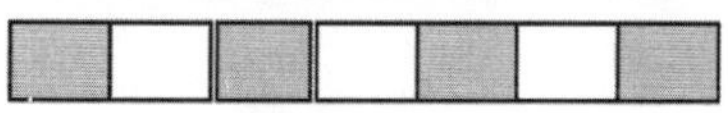
d. 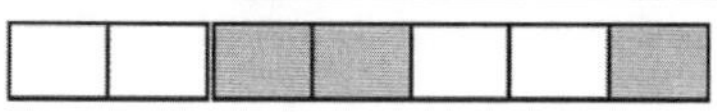

12. The number of miles Jacob has walked each year over the past five years is shown in the table below.

Year 1	691
Year 2	567
Year 3	144
Year 4	963
Year 5	221

Which sequence of years shows the number of miles he walked each year, listed in order from greatest to least?

a. Year 2, Year 4, Year 5, Year 1, Year 3
b. Year 4, Year 1, Year 2, Year 5, Year 3
c. Year 1, Year 4, Year 2, Year 5, Year 3
d. Year 3, Year 2, Year 5, Year 4, Year 1

13. A neighborhood contains 3 streets. Street 1 has 26 houses, Street 2 has 12 houses, and Street 3 has 35 houses. Estimate by rounding, how many houses in total are on Street 1 and Street 3?

a. 30
b. 40
c. 50
d. 60

14. A little girl has two nickels, two dimes, four pennies, and one quarter. How much money does she have?

a. 30 cents
b. 90 cents
c. 60 cents
d. 40 cents

15. What is the seventh number in the pattern below?

228, 235, 242, 249, ___, ___, ___

a. 269
b. 256
c. 270
d. 263

16. A farmer plants rows of corn each growing season. The table below shows the total number of rows of corn the farmer has planted after several seasons.

Season	Number of Rows
Season 2	34
Season 4	68
Season 6	102
Season 8	136
Season 10	

How many rows of corn will the farmer have planted by the end of Season 10?

a. 153
b. 170
c. 136
d. 168

17. In which place value is the digit 4 in 7,843?

a. Ones
b. Tens
c. Hundreds
d. Thousands

18. Anand has 45 stamps in his collection. He decides to give all of his stamps to 5 friends. He writes the number sentence below to find out how many stamps to give each friend, if each friend is to receive an equal amount.

$$45 \div 5 = ?$$

Which number sentence would NOT help him find the number of stamps to give each friend?

a. $? \times 5 = 45$
b. $45 - 5 = ?$
c. $45 \div ? = 5$
d. $5 \times ? = 45$

19. What is the next item in the pattern below?

a. 

b. 

c. 

d. 

20. Hannah builds the towers shown below, using square blocks.

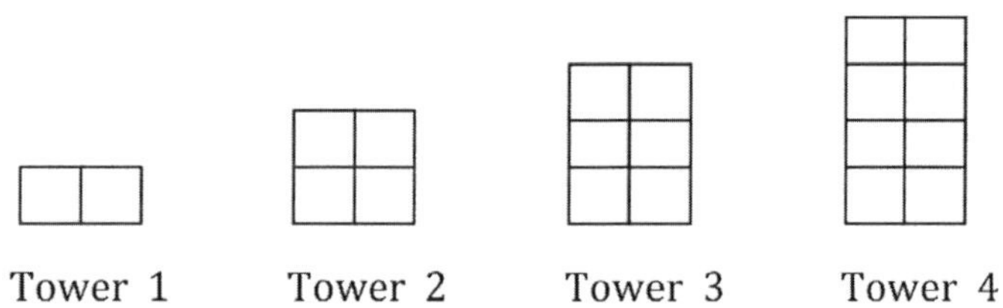

If she continues this pattern, how many square blocks will she use in the eighth tower?

a. 12
b. 16
c. 18
d. 14

21. The table below shows the number of books students brought to share with the class.

Number of Students	Number of Books
2	8
5	20
6	24
8	32

How many books did 12 students bring?

a. 44
b. 36
c. 40
d. 48

22. Jordan wants to show the number of wheels found on several cars. Which table should he use?

a.

Number of Cars	Number of Wheels
3	12
4	16
7	28
11	44
12	48

b.

Number of Cars	Number of Wheels
2	6
5	15
8	24
9	27
12	36

c.

Number of Cars	Number of Wheels
3	15
5	25
6	30
8	40
11	55

d.

Number of Cars	Number of Wheels
4	8
6	12
9	18
12	24
14	28

23. What is 49,249 rounded to the nearest hundred?

a. 49,200
b. 200
c. 49,250
d. 49,300

24. Which operation symbol properly compares $\frac{1}{3}$ and $\frac{11}{12}$?

a. <
b. =
c. >
d. Impossible to determine

25. Which number is greater than the number shown by Point S on the number line?

a. $14\frac{1}{4}$
b. 14
c. $14\frac{3}{4}$
d. $14\frac{1}{2}$

26. Which of the following values lies between $\frac{3}{14}$ and $\frac{7}{14}$?

a. $\frac{1}{14}$
b. $\frac{3}{14}$
c. $\frac{9}{14}$
d. $\frac{6}{14}$

27. Which of the following is congruent to the triangle shown below?

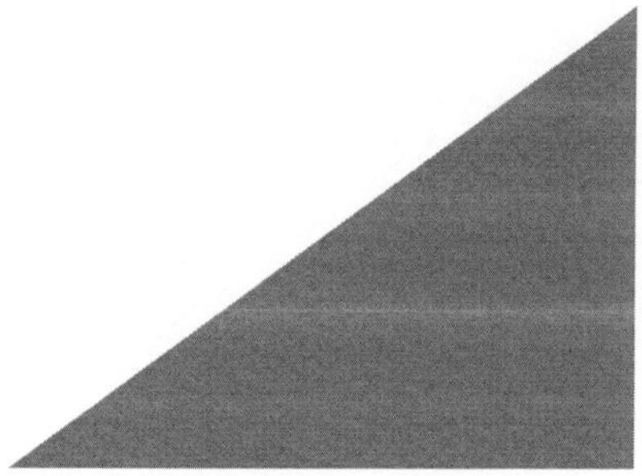

a.  c. 

b. 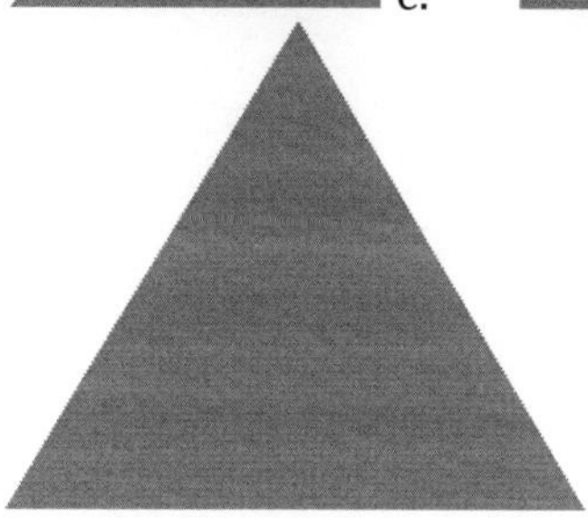 d. 

28. Which of the following figures has fewer than 6 sides?

a. Pentagon
b. Octagon
c. Nonagon
d. Decagon

29. What is 6,709 rounded to the nearest ten?

a. 6,700
b. 6,710
c. 7,000
d. 6,810

30. Ashley states, "All two-dimensional shapes are polygons." Which of the following shapes proves this statement to be false?

a. Triangles
b. Rectangles
c. Circles
d. Hexagons

31. What is the perimeter of the shape shown below?

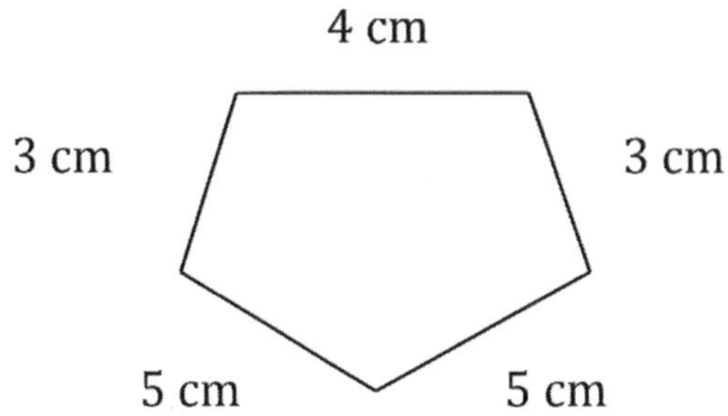

a. 29 cm
b. 20 cm
c. 15 cm
d. 25 cm

32. Which of the following is true?

a. $\frac{7}{12} = \frac{9}{12}$
b. $\frac{9}{12} < \frac{9}{12}$
c. $\frac{9}{12} > \frac{7}{12}$
d. $\frac{9}{12} = \frac{7}{12}$

33. About how wide is the drawing of the notebook shown below?

a. 2 cm
b. 3 cm
c. 5 cm
d. 7 cm

34. How many square units are found in the shape below?

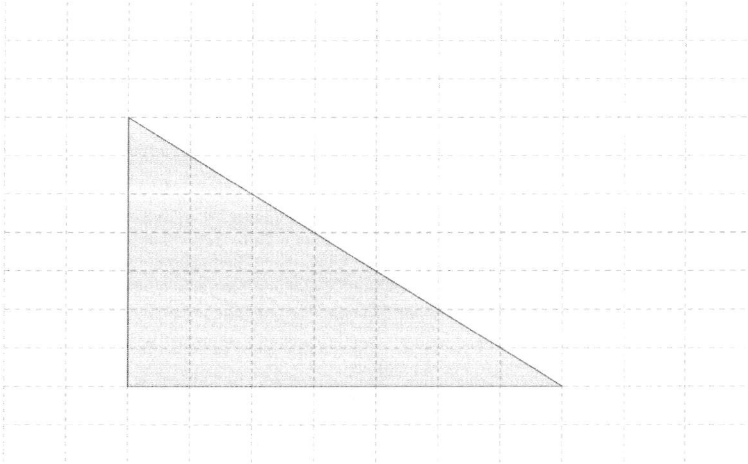

a. 23 square units
b. $23\frac{1}{2}$ square units
c. 24 square units
d. $24\frac{1}{2}$ square units

35. Which operation symbol properly compares $\frac{2}{4}$ and $\frac{5}{10}$?

a. <
b. =
c. >
d. Impossible to determine

36. The outside temperature, on a spring day, in Flagstaff, Arizona, is shown on the thermometer below.

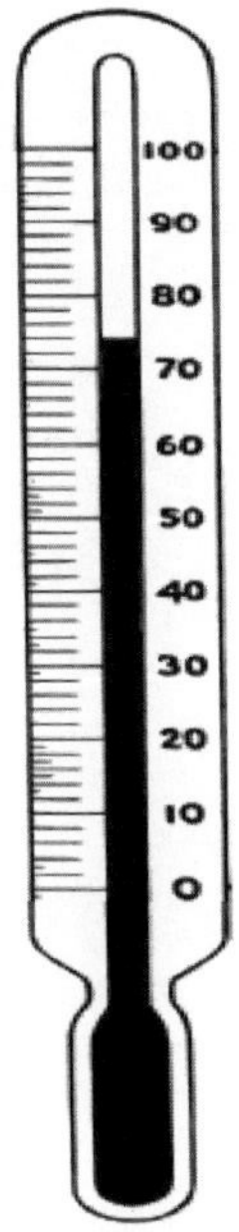

What is the temperature, in degrees Fahrenheit?

a. 77° F
b. 76° F
c. 79° F
d. 74° F

37. Amanda arrives at a birthday party at the time shown on the clock below.

What time did she arrive at the party?

a. 10:10
b. 2:10
c. 9:10
d. 2:50

38. Which clock below would be read as a quarter past three?

a.

3:25

b.

3:15

c.

3:45

d.

3:30

39. A bag contains 3 red cards, 7 blue cards, 9 green cards, and 6 yellow cards. Jesse randomly draws a card from the bag. Which of the following statements is true?

a. He is less likely to draw a green card than a yellow card
b. He is more likely to draw a yellow card than a red card
c. He is more likely to draw a yellow card than a blue card
d. He is equally likely to draw a red, blue, green, or yellow card

40. The number of cakes baked in one month, by different bakeries, is shown in the table below.

Bakery	Bakery A	Bakery B	Bakery C	Bakery D
Number of Cakes Baked	25	15	35	20

Which pictograph shows the number of cakes baked by each bakery?

a. Each picture of a cake represents 5 cakes.

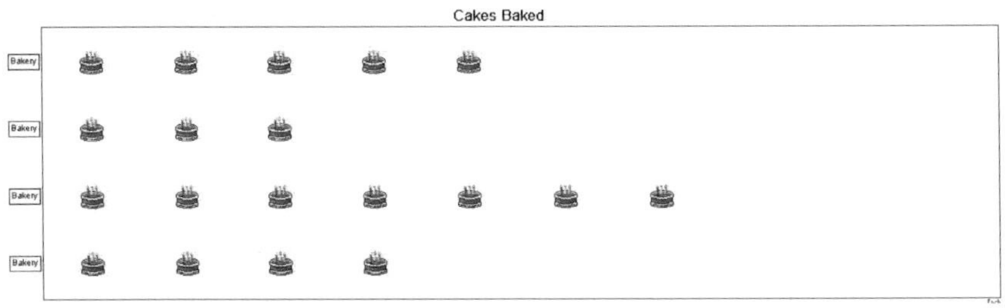

b. Each picture of a cake represents 4 cakes.

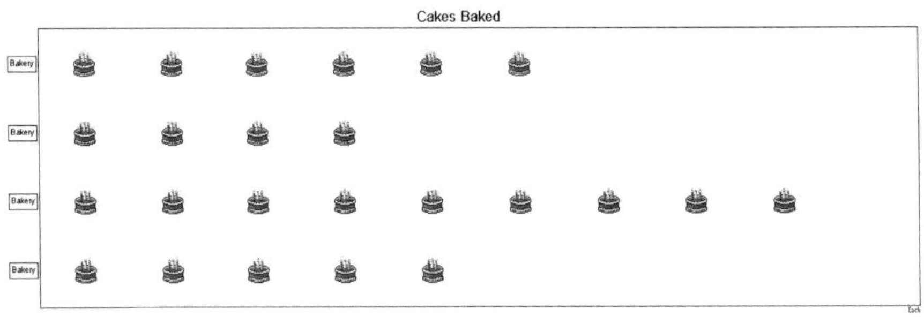

c. Each picture of a cake represents 3 cakes.

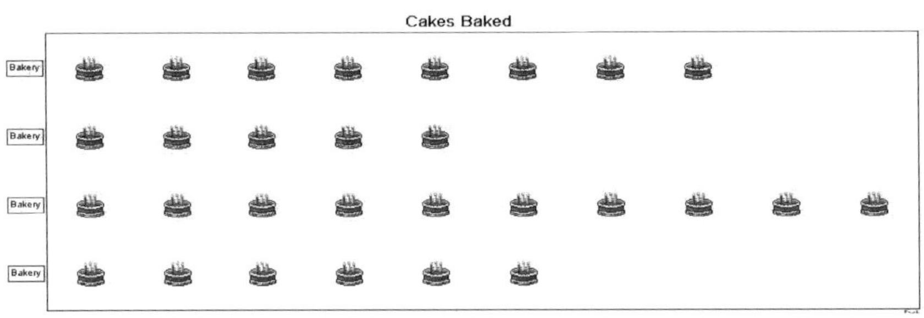

d. Each picture of a cake represents 6 cakes.

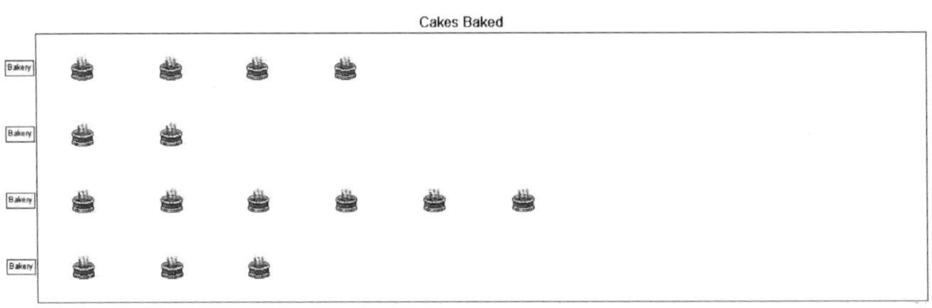

41. The bar graph below shows the number of Math teachers, from four different states, attending a math event.

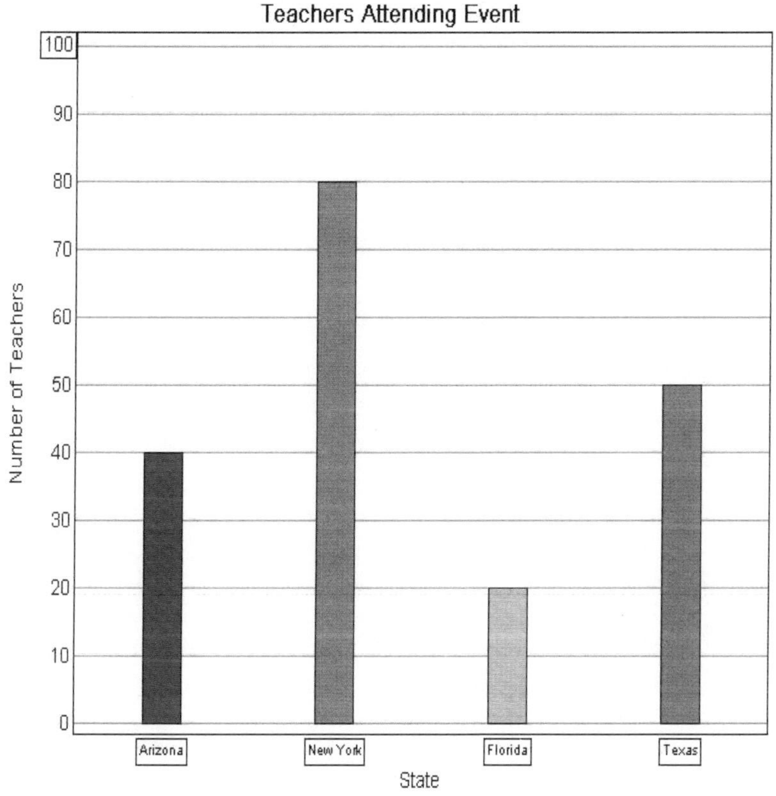

Which state has the fewest number of teachers attending the math event?

a. Arizona
b. New York
c. Florida
d. Texas

42. A student spins a spinner, with sections, labeled 1 – 8. Which of the following best represents the likelihood of the spinner landing on a 9?

a. Likely
b. Not likely
c. Certain
d. Impossible

43. Order these numbers from greatest to least.

591	602	592	573	584	588

a. 602; 592; 591; 584; 588; 573
b. 602; 591; 592; 588; 584; 573
c. 573; 584; 588; 591; 592; 602
d. 602; 592; 591; 588; 584; 573

44. **The bar graph below shows the number of votes for choosing a class mascot.**

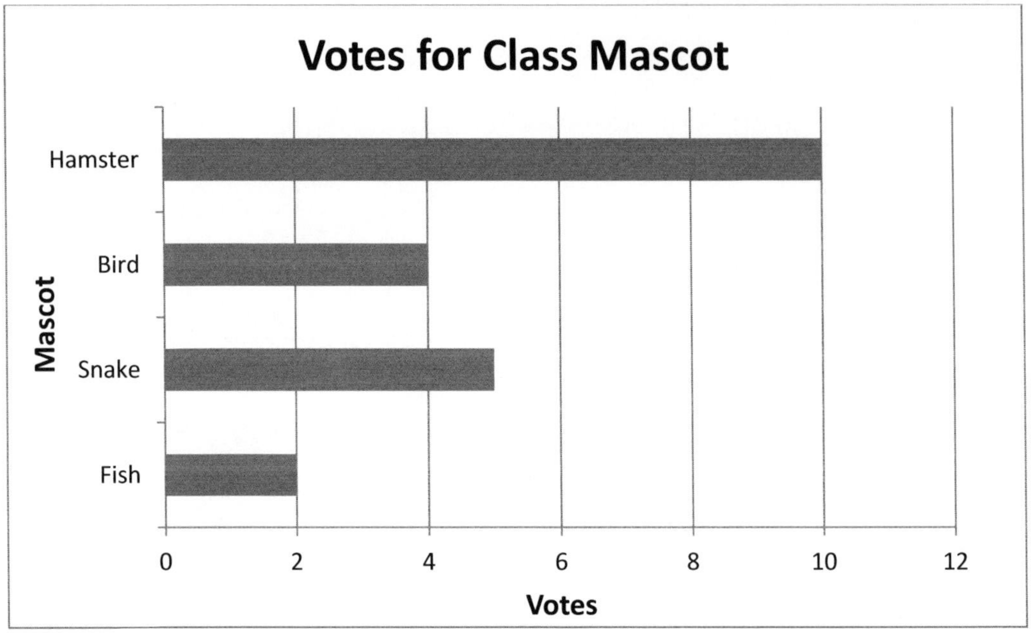

Solve and place answer on the provided griddable sheet.

How many more votes did the hamster receive than the fish?

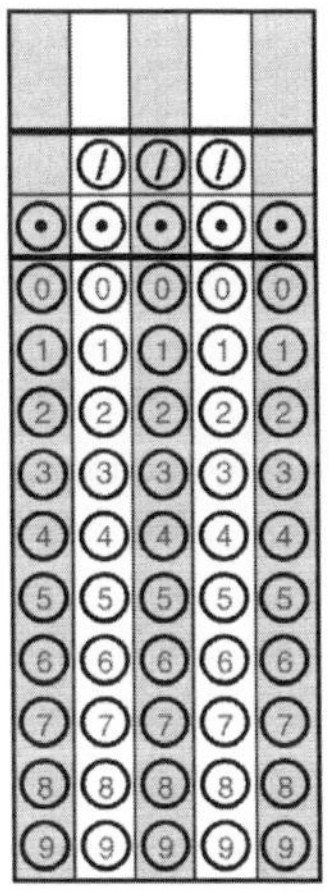

Answer Key and Explanations for Test #2

1. A: The place value is assigned by the position of the digit. The places from left to right are ones, tens, hundreds, thousands, ten-thousands, etc. Since the digit 5 is in the fourth place from the left, it is in the thousands place. The value of the digit is 5×1000, which is 5000.

2. D: All numbers for Choice D have a 5 in the hundreds place, indicating that the next digit, in the tens place, must be used to make comparisons. Each digit in the tens place is different, so an examination of the ones place is not needed. The list of increasing tens place digits for Choice D are 2, 3, 4, 7, and 8, indicating a correct ordering of the numbers from least to greatest. None of the other choices show strictly increasing numbers.

3. C: The term subtracted from means to "take away" and place a minus sign in the equation. Since 5 is being subtracted from 25, write the number 25 first followed by a minus sign and then the number 5. The problem states that this subtraction equals 5, so simply attach this to the end.

4. D: Choice D shows 1 crayon and 3 pencils, indicating that 1 out of 4 of Camille's writing tools were crayons. The other choices are out of either a total of 7 writing tools or 5 writing tools, which cannot be used in a ratio that represents 1 out of 4.

5. D: In order to find out the total number of hot dogs she sold in the two months, the amount sold in each month should be added. The sum of 128 and 117 is 245. Thus, she sold 245 hot dogs during the two months.

6. D: Two dollar bills are worth \$2.00. A quarter is worth \$0.25, 4 dimes are worth \$0.40 since they are worth \$0.10 each ($4 \times 0.10 = 0.40$), 3 nickels are worth \$0.15 since they are worth \$0.05 each ($3 \times 0.05 = 0.15$), and 2 pennies are worth \$0.02 since each is worth \$0.01. The sum of the coins is: \$0.25 + \$0.40 + \$0.15 + \$0.02, which equals \$0.82. The sum of the two dollar bills and the coins can be written as: \$2.00 + \$0.82, which equals \$2.82.

7. D: The diagram shows 24 counters total, divided into 3 groups, with 8 counters in each group. Therefore, the total number of counters, 24, is divided by 3. This gives a quotient of 8, which is written as: $24 \div 3 = 8$.

8. 312: This is a simple multiplication problem which involves carrying the excess (greater than a single digit) to the next column. Start with the 4 and multiply each digit above: $4 \times 8 = 32$. Write down the 2 and place the 3 on top of the 7. Multiply $4 \times 7$to get 28. Add in the 3 to get 31. Write down the number 31 and the product is 312.

9. B: The number of customers who entered the store can be rounded to 90. That number can be multiplied by 3, since each customer receives 6 coupons. The product of 90×3 is 270.

10. A: Since a chicken has 2 feet and a horse has 4 feet, you may multiply the number of chickens found on each farm by 2, and then multiply the number of horses found on each farm by 4. After finding the total number of chicken feet and horse feet on each farm, you can then add the two amounts to find the total number of animal feet on each farm. You can then compare the two values to determine the one that is larger.

11. B: Each model shows a whole, split into 7 sections. This makes the denominator equal to 7 for the fraction representing the shaded section for each. Since the fractions all have the same number on the bottom, the number of shaded sections can be compared. For a fraction to be more than $\frac{4}{7}$, more than 4 parts must be shaded. The model shown for Choice B shows the fraction $\frac{6}{7}$, because 6 out of 7 sections are shaded, showing a fraction that is more than 4 out of 7. The other choices all show either 4 parts or less than 4 parts shaded.

12. B: The sequence of the number of miles walked in order from greatest to least is: 963 (Year 4), 691 (Year 1), 567 (Year 2), 221 (Year 5), 144 (Year 3). The number of miles walked can also be compared by examining only the digits in the hundreds place since they are all different. 9 is greater than 6 which is greater than 5 and so on. The only choice that shows the years for the number of miles walked in descending order is B.

13. D: $26 + 35 = 61$. Since the 1 in the ones digit is less than five, the number is rounded down to 60.

14. C: Convert the coins to cents and apply addition. Each nickel is worth 5 cents, so 2 nickels equals 10 cents. Each dime is worth 10 cents, so 2 dimes equals 20 cents. Each penny is worth 1 cent, so 4 pennies equals 4 cents. A quarter is worth 25 cents. $10 + 20 + 4 + 25 = 59$ cents which can be rounded to 60 cents to match one of the provided answer selections.

15. C: Each number is 7 more than the previous number, so adding 7 to the previous number finds the next number. Thus, the fifth number is 256, the sixth number is 263, and the seventh number is 270.

16. B: The farmer plants 17 rows of corn each season. This can be found using all of the information given in the table. The farmer had planted 34 rows of corn by the end of Season 2 and 68 rows of corn by the end of Season 4, indicating an increase of 34 rows of corn between the two seasons. So, by dividing 34 by 2, the numbers of rows planted in one season is found. Also, the farmer had planted 102 rows of corn by the end of Season 6 and 136 rows of corn by the end of Season 8, indicating an increase of 17 rows of corn planted in one season. If 17 rows of corn are added to the number given at the end of Season 2, the result is 51 rows of corn planted by the end of Season 3. If another 17 rows of corn are added to this amount, the farmer would have planted 68 ears of corn by the end of Season 4, which he did. Thus, he did plant 17 rows of corn each year. He had planted 119 rows of corn by the end of Season 7, 153 rows of corn by the end of Season 9, and 170 rows of corn by the end of Season 10.

17. B: The place value is assigned by the position of the digit. The places from left to right are ones, tens, hundreds, thousands, ten-thousands, etc. The 4 is located in the tens place since it is two spaces from the left.

18. B: A number sentence that subtracts the number of friends from the total number of stamps will not provide the number of stamps needed to give each friend. Instead, an appropriate multiplication or division number sentence within the following fact family is needed: $9 \times 5 = 45$, $5 \times 9 = 45$, $45 \div 9 = 5$, or $45 \div 5 = 9$.

19. D: The pattern displayed is an arrow pointing right, left, up, right, left....so the next arrow should be pointing up to continue the pattern.

20. B: The number of blocks used in each tower increases by 2. Therefore, the number of blocks used in the next four towers can be found by adding 2 to the number of blocks in the 4th tower and continuing to add 2 blocks for each tower that comes next. This gives: 10, 12, 14, and 16, with 16 blocks used in the eighth tower.

21. D: The number of books brought by 2 students is 8, while the number of books brought by 5 students and 6 students increased by 4. Thus, the number of books brought by each student was 4. This fact can be checked by starting with 1 student and 4 books brought, and continuing the pattern to make sure it corresponds with the numbers in the table. For example, 8 students brought 32, which does in fact agree with the each student bringing 4 books. So, the number of books each student brought, 4, is multiplied by the number of students to find the total number of books that were brought. Thus, 12 students brought 12×4 books, or 48 books.

22. A: Each car has 4 wheels, so the numbers of cars should be multiplied by 4 to find the total number of wheels. 3 cars will have 3×4 wheels, or 12 wheels, 4 cars will have 4×4 wheels, or 16 wheels, 7 cars will have 7×4 wheels, or 28 wheels, 11 cars will have 11×4 wheels, or 44 wheels, and 12 cars will have 12×4 wheels, or 48 wheels.

23. A: The place value of a digit depends on its location in relation to the decimal point. The places to the left of the decimal are ones, tens, then hundreds, respectively. To round a number to a certain place value, look at the digit to the immediate right. If the digit to the right is 5 or more, round up by adding one followed by a zero. If the digit is less than five, round down by keeping the same digit followed by a zero. For this problem, the digit in the hundreds place is 2. The digit to its immediate right is 4. Since 4 is less than 5, round down. So, 49,249 rounded to the nearest hundred is 49,200.

24. A: One way to compare two fractions is by making two strip diagrams. The denominator represents how many equal spaces the strip will be divided into. The numerator represents how many spaces on the strip will be shaded in. It is important to make the strip diagrams the same size to accurately compare the fractions. When looking at the strip diagrams, it is evident that $\frac{1}{3}$ is less than $\frac{11}{12}$.

$\frac{1}{3}$

$\frac{11}{12}$

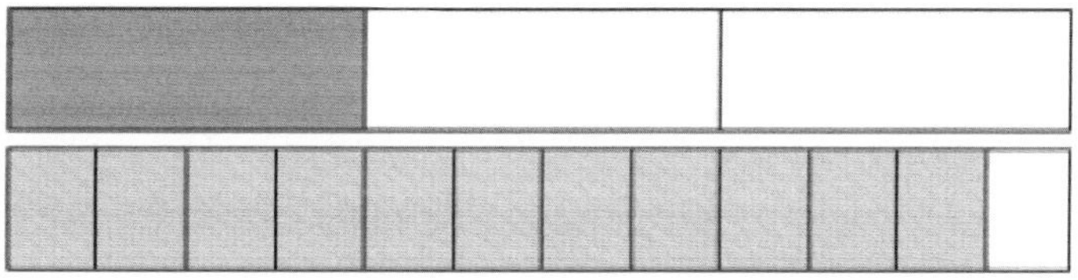

25. C: The number line shows three marks between each whole number, making 4 equal parts between each whole number. Point S is located at $14\frac{1}{2}$ since it is halfway between 14 and 15, or 2 marks to the right of 14, which can also be written as $14\frac{2}{4}$. $14\frac{3}{4}$ is greater than that number. It is easier to see which fraction is greater if the denominators are the same, since then, just the numerators are compared. So, $14\frac{3}{4}$ is $\frac{1}{4}$ more than the number shown on the number line, and this is the only choice that is greater than the value of Point S.

26. D: $\frac{6}{14}$ is the only value that comes between $\frac{3}{14}$ and $\frac{7}{14}$ on the number line. $\frac{1}{14}$ is less than $\frac{3}{14}$, so it is before $\frac{3}{14}$ on the number line. $\frac{3}{14}$ is equal to $\frac{3}{14}$, so it is at the exact place on the number line. $\frac{9}{14}$ is greater than $\frac{3}{14}$, so it is after $\frac{3}{14}$ on the number line.

27. C: The triangle shown for Choice C is congruent to the given triangle, since all side lengths and angles are congruent. Basically, the triangles must be the same size to be congruent, but can be flipped or rotated in any way.

28. A: A pentagon has 5 sides. An octagon has 8 sides. A nonagon has 9 sides. A decagon has 10 sides.

29. B: The place value of a digit depends on its location in relation to the decimal point. The places to the left of the decimal are ones, tens, then hundreds, respectively. To round a number to a certain place value, you look at the digit to the immediate right. If the digit to the right is 5 or more, round up by adding one. If the digit is less than five, round down by keeping the same digit. For this problem, the digit in the tens place is 0. The digit to its immediate right is 9. Since 9 is greater than 5, round up. So, 6,709 rounded to the nearest ten is **6,710.**

30. C: A circle is a two-dimensional shape that is not a polygon. It does not have straight sides that meet at vertices. All of the other choices are two-dimensional polygons.

31. B: The perimeter is the sum of the lengths of all five sides, or $5 + 5 + 3 + 3 + 4$, which equals 20. Therefore, the perimeter of the pentagon is 20 cm.

32. C: $\frac{9}{12}$ is greater than $\frac{7}{12}$. They are not equal. When comparing fractions with the same denominator, compare only the values in the numerator. Since they share a common denominator of 12, compare only the 9 and 7. Since 9 is greater than 7, the same can be said for the fraction $\frac{9}{12}$ when compared to $\frac{7}{12}$.

33. C: The width of the notebook can be estimated by marking off lengths that are close to the measurement of a centimeter. The width of the book is slightly less than 5.5 cm, so it is about 5 cm.

34. D: The triangle includes 21 whole square units, plus 7 one-half square units, or $3\frac{1}{2}$ square units. So, 21 square units plus $3\frac{1}{2}$ square units gives a total of $24\frac{1}{2}$ square units.

35. B: One way to compare two fractions is by making two strip diagrams. The denominator represents how many equal spaces the strip will be divided into. The numerator represents how many spaces on the strip will be shaded in. It is important to make the strip diagrams the same size to accurately compare the fractions. When looking at the strip diagrams, it is evident that $\frac{2}{4}$ is equal to $\frac{5}{10}$.

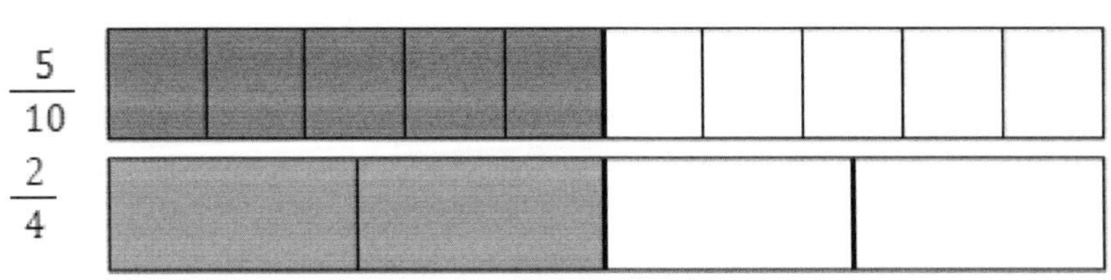

36. D: The thermometer shows 4 marks between each whole number, or 5 intervals. This means each interval on the thermometer represents 2 degrees since there are 5 intervals between each difference of 10 degrees. The thermometer reveals a reading at 4 degrees above 70 degrees (2 marks above 70), or 6 degrees below 80 degrees (3 marks below 80). Thus, the temperature outside is 74 degrees Fahrenheit.

37. A: The short hand, or hour hand, is between 10 o'clock and 11 o'clock, revealing that Amanda arrived after 10 o'clock, but before 11 o'clock. It is much closer to the 10, so this indicates the time is much closer to 10 o'clock than 11 o'clock. The long hand, or minute hand, is pointing to the 2, indicating 10 minutes after the hour. This is because for minutes, each number represents 5 minutes; so $2 \times 5 = 10$. So, she arrived at the party at 10:10.

38. B: Since there are sixty minutes in an hour, a "quarter" in time represents 15 minutes. A quarter past three would mean 15 minutes after 3 o'clock or 3:15. Represented on an analog clock, the big hand would be pointing to the three and the small hand pointing just past the three.

39. B: There are 6 yellow cards and 3 red cards. The more cards there are of a certain color, the more likely it is that the color is drawn. With more yellow cards than red cards in the bag, he is more likely to draw a yellow card than a red card. There are more green cards than yellow cards, so he is more likely to draw a green card than a yellow card. There are less yellow cards than blue cards, so he is less likely to draw a yellow card than a blue card. Finally, the number of red, blue, green, and yellow cards are all different – so none of them are equally likely to be drawn compared to another color. Thus, Choice B is the only true statement.

40. A: The pictograph for Choice A reveals that each picture of a cake represents 5 actual cakes. Therefore, the numbers given in the table, divided by 5, should equal the number of cakes represented in the pictograph. The pictograph accurately shows 5 cakes for 25 actual cakes, 3 cakes for 15 actual cakes, 7 cakes for 35 actual cakes, and 4 cakes for 20 actual cakes. This is the only pictograph that represents the correct number of cakes.

41. C: Florida had 20 teachers that attended the event, which is less than the number of teachers who attended the event from each of the other three states. The number of teachers that attended the event from each of the other states were 40 (Arizona), 80 (New York), and 50 (California). Also, just looking at the bar graph shows that Florida had the least number of teachers attend compared to the other states because the bar is much lower in the graph.

42. D: Since the spinner only has sections, labeled 1-8, there is not a section, labeled "9". Therefore, it is impossible for the spinner to land on a 9.

43. D: The correct order when order from greatest to least is 602; 592; 591; 588; 584; 573. 602 is the furthest to the right on the number line, so it is the largest. 573 is the furthest to the left on the number line, so it is the smallest.

44. 8: The hamster received 10 votes and the fish received 2 votes. The difference is $10 - 2 = 8$ votes.

Online Resources

Due to our efforts to try to keep this book to a manageable length, we've created a link that will give you access to all of your online resources:

mometrix.com/resources719/vsolg3math

It's Your Moment, Let's Celebrate It!

Share your story @mometrixtestpreparation